AEROSOLS
In the Mining and Industrial Work Environments

VOLUME 1
Fundamentals and Status

Edited by

Virgil A. Marple
Benjamin Y. H. Liu

ANN ARBOR SCIENCE
THE BUTTERWORTH GROUP

PREFACE

These volumes contain the papers presented at the International Symposium on Aerosols in the Mining and Industrial Work Environment held in Minneapolis, Minnesota, November 1-6, 1981. Included in these volumes is a total of 81 papers, with authors from 10 countries.

The objective of the symposium, and hence of these volumes, was to summarize the state of knowledge in the field of work-environment aerosols and to provide a forum at which the most recent findings and important new developments could be presented. The symposium succeeded largely because it attracted the leading experts in the field from most of the industrialized nations of the West and Japan.

Much of the interest in work-environment aerosols has derived from the realization in recent years that airborne particles in the workplace environment often contain unhealthful, and sometimes toxic, components. The influence of these substances on the health of the workers often does not manifest itself for many years—sometimes for periods exceeding 10, 20, 30 or more years. Familiar examples include coal, silica, asbestos and other mineral dusts, cotton and wood dusts, and airborne radioactive particles, to name just a few. The purpose of these volumes is not to address any one of these problems in detail; the focus is on the common fundamental problems: instrumentation, sampling strategies, standards, calibration, etc. Only by paying particular attention to these fundamental problems can valid results be obtained.

For the purpose of organizing the material, the symposium papers are divided into nine main parts, which are in turn organized into three separate volumes. Part 1 provides a general status review on work-environment aerosols in various countries, prepared by leading experts from their respective countries. In Part 2, the interaction between particles and the human respiratory system is considered. In Part 3, the questions of sampling strategies and the efficiency of sampling inlets are addressed. Part 4 deals with aerosols in the mining work environment; Part 5 deals with aerosols in the general industrial work environment. In Part 6, the problem of asbestos is considered. This is a particularly important problem, not only because of the health effects of asbestos, but because of the special measurement problems associated with asbestos particles. Because of the large number of papers on instrumentation presented at the symposium, the instrumentation papers are divided into the final three parts, which constitute Volume 3.

V. A. Marple
B. Y. H. Liu

ACKNOWLEDGMENTS

A large undertaking such as the present one cannot succeed without the help of many individuals. We wish to thank Mr. Kenneth L. Williams, of the Bureau of Mines, and Dr. Paul A. Baron, of the National Institute for Occupational Safety and Health, for helping to secure the necessary financial support for the symposium and for their general guidance and counsel, including their service on the symposium Program Committee. We also wish to acknowledge other members of the Program Committee (PC), all of whom provided valuable input, and the help of the Session Chairmen (SC) in conducting the sessions:

- Michel M. Benarie, Institut National de Recherche Chimique Appliquee, France (PC, SC);
- Melvin E. Cassady, Occupational Safety and Health Administration Laboratory, Utah (PC);
- C. Norman Davies, University of Essex, Great Britain (PC);
- Heinz J. Fissan, Universität Duisberg, Federal Republic of Germany (SC);
- Katsunori Homma, National Institute of Industrial Health, Japan (PC, SC);
- Walter John, Air and Industrial Hygiene Laboratory, California (PC, SC);
- David B. Kittleson, University of Minnesota, Minnesota (SC);
- Geoffrey Knight, Elliot Lake Laboratory, Canada (PC, SC);
- Michael J. Larsen, Occupational Safety and Health Administration Health Response Team, Utah (PC, SC);
- Morton Lippmann, New York University Institute of Environmental Medicine, New York (PC, SC);
- Peter H. McMurry, University of Minnesota, Minnesota (SC);
- Trevor L. Ogden, Occupational Medicine and Hygiene Laboratories, Great Britain (SC);
- Lars Olander, National Board of Occupational Safety and Health, Sweden (PC, SC);
- Vittorio Prodi, Laboratorio Fisica Sanitoria, Italy (PC, SC);
- George H. Schnakenberg, Jr., U.S. Bureau of Mines, Pennsylvania (PC);
- Werner Stöber, Fraunhofer Institute for Toxicology and Aerosol Research, Federal Republic of Germany (PC, SC);
- Thomas F. Tomb, U.S. Mine, Safety and Health Administration, Pennsylvania (PC, SC).
- Jon Volkwein, U.S. Bureau of Mines, Pennsylvania (PC, SC); and
- Robert W. Wheeler, National Institute for Occupational Safety and Health, West Virginia (PC, SC).

We are also thankful to Mr. Joseph M. Kroll and Ms. Moira A. Keane, of the Department of Conferences of the University of Minnesota, for helping to organize the symposium and taking care of the myriad organizational details inherent in a meeting of this type. We wish to thank also members of the Particle Technology Laboratory staff and students at the University of Minnesota who helped to run the symposium. Particular thanks go to Dr. David Y. H. Pui, the manager of the laboratory for helping to manage many important aspects of the symposium and for helping to handle and process the manuscripts. The help of Dr. Kenneth L. Rubow, also of the Particle Technology Laboratory, in undertaking the difficult and time consuming task of compiling the subject index for these volumes is gratefully acknowledged. Finally, we wish to thank Ms. Linda Stoltz for her help in typing, in keeping the line of communication to the authors open and in taking care of numerous other matters relating to the compilation and editing of these volumes.

Marple   Liu

Virgil A. Marple is Associate Professor of Mechanical Engineering at the University of Minnesota. Dr. Marple holds BME and PhD degrees from the University of Minnesota and an MS degree from the University of Southern California, all in the field of Mechanical Engineering. He was an engineer with several industrial and aerospace companies before joining the University of Minnesota. Currently he is active in teaching and research in particle technology and aerosol science. His research work has been primarily in instrumentation and has included the development of numerical techniques for modeling flow fields and particle trajectories in inertial separators, and the development of aerosol monitoring instruments, and techniques, and equipment for instrument calibration. Dr. Marple is the author of numerous technical papers and is active in many professional societies.

Benjamin Y. H. Liu is Professor of Mechanical Engineering and Director of the Particle Technology Laboratory at the University of Minnesota. Dr. Liu is a leading authority in the field of aerosol science and engineering. He has served as a visiting professor at the University of Paris VI, a visiting scientist at the Oak Ridge National Laboratory, a member of the Particulate Control Technology Committee of the National Academy of Sciences, a Guggenheim Fellow, a member of the editorial advisory boards of the *Journal of Aerosol Science* and the *Journal of Colloid and Interface Science*, and as a technical consultant to various industrial companies. He is best known for his basic research on electrical charging and precipitation, aerosol sampling and transport, filtration, and inertial impaction. He is the developer or the co-developer of several widely used aerosol generating, measuring and sampling instruments, including the vibrating-orifice monodisperse aerosol generator, the pulse-precipitating electrostatic aerosol sampler, the electrical mobility analyzer and classifier, and the constant output atomizer. He is the editor of *Fine Particles: Aerosol Generation, Measurement, Sampling, and Analysis*, and the editor-in-chief of the journal *Aerosol Science and Technology*. He is also a fellow of the American Society of Mechanical Engineers and the American Association for the Advancement of Science.

CONTENTS

Part 1
Status of Work-Environment Aerosols

Part 2
Particles and the Respiratory System

Part 3
Sampling Strategy, Data Analysis and Inlet Efficiency

PART 1

STATUS OF WORK-ENVIRONMENT AEROSOLS

CHAPTER 1

STATUS OF WORK-ENVIRONMENT AEROSOLS
IN GREAT BRITAIN

R. J. Hamilton

    Mining Research and Development Establishment
    National Coal Board
    Burton-on-Trent, United Kingdom

T. L. Ogden

    Occupational Medicine Hygiene Laboratory
    Health and Safety Executive
    London, United Kingdom

J. H. Vincent

    Institute of Occupational Medicine
    Edinburgh, United Kingdom

ABSTRACT

The organization of government control and health standards in Great Britain is described, together with the present position with respect to dust sampling techniques and the special problems posed by the measurement of "total" dust and fibers. In addition to the work carried out in government departments and institutes, there is a tradition of research and development in the "dusty" industries themselves. The history of the successful struggle against coalworker's pneumoconiosis is described, with reference to research into dust measurement, an epidemiological study covering all British coalfields, and the considerable improvement in dust control techniques. Finally, the effects of these efforts on the disease are illustrated. The chapter ends with a discussion of present work on aerosol problems in different industries, and gives a brief description of requirements for future work.

INTRODUCTION

Britain was the first country to experience an industrial revolution, and has had its share of the diseases associated with dusty industries, including silicosis, coalworker's pneumoconiosis, byssinosis and asbestosis. Various organizations have been involved in a long history of aerosol research related to health problems. The Chemical Defence Experimental Establishment, Porton, was responsible for many early instrumental developments, including the thermal precipitator (TP) conifuge and cascade impactor. The Medical Research Council (MRC) has a long association with the subject, and the MRC Pneumoconiosis Unit, Cardiff, had its origin in the classical studies in the South Wales coalfield. The Safety in Mines Research Establishment, which is now part of the Health and Safety Executive (HSE), was originally founded with the main aim of solving the problems of dust explosions in mines, and has a long history of successful research and development in dust physics. More recently, the Institute of Occupational Medicine (IOM), Edinburgh, which developed out of the National Coal Board (NCB) Pneumoconiosis Field Research, has contributed to work on both coal-mining dust problems and on those connected with asbestos and other fibers.

In addition to the work of the various institutes and universities, the dusty industries themselves have played an important role in advancing aerosol studies: the British Ceramic Research Association and the British Cast Iron Research Association, for example, have made important contributions in basic studies and applied dust control in their respective industries. Perhaps the history of airborne dust in coal mining over the last 40 years or so demonstrates the best known example of this type of work. This chapter covers some of the main themes of research with work environment aerosols, with special reference to the coal industry, and tries also to look into the future.

In 1974 the Health and Safety at Work Act set up the Health and Safety Commission (HSC), which was to assist and encourage securing the health, safety and welfare of workers; the protection of others against risks to health and safety arising from work processes; and control of dangerous substances and emissions. It has a responsibility for research and training, and proposes regulations to the government. The HSC is a committee of nine, representing employers, employees and other organizations, appointed by the Secretary of State for Employment. It has permanent advisory committees, such as the Advisory Committee on Toxic Substances, responsible for much of the work on new standards, and temporary committees, such as the Advisory Committee on Asbestos, which recently advised the government on new standards and controls for these materials [HSC 1979a].

The HSE is responsible to the HSC; HSE includes the Factory Inspectorate, the Mines and Quarries Inspectorate, the Alkali Inspectorate (which deals with emissions from certain industrial processes), the Employment Medical Advisory Service, resources and policy formulating divisions, and the Research and Laboratory Services Division (which provides laboratory support for the other departments).

STANDARDS FOR AIRBORNE CONCENTRATIONS

HSC policy on airborne materials is described in two guidance notes [HSE 1978,1981a]. The attitude to standards is that "exposure should be kept as low as is reasonably practicable" and "in any case exposure should be kept within the published standards." "Where there are no known toxic effects, there should still be a policy of keeping exposure as low as is reasonably practicable." The published standards include a list of threshold limit values (TLV) [HSE 1981a]. This is a reprint of the American Conference of Government Industrial Hygienists (ACGIH) list, with an introduction explaining HSC policy, and giving substances for which HSC has a standard different from that in the ACGIH list. Some of these standards are control limits, i.e., levels agreed by the HSC, but not necessarily meeting the ACGIH definition of TLV as a level to which nearly all workers may be exposed repeatedly without adverse effect. Where exposures cannot be kept below the standard, personal protective equipment should be used, but source control is preferred, and the requirement is emphasized to keep levels as low as reasonably practicable, and not simply below the TLV.

RESPIRABLE DUST

Not surprisingly, Britain uses the definition of respirable dust adopted in 1952 by the MRC, described by Hamilton and Walton [1961]. A variety of horizontal elutriator instruments and personal samplers using the respirable dust selecting cyclone that follows the MRC curve are in use [Higgins and Dewell 1966; Maguire et al. 1973].

In Britain, the ACGIH TLV for crystalline silica are applied using the MRC respirable definition. However, the difference between measurements obtained using the ACGIH and MRC definitions of respirability is usually small compared with the discrepancies produced by other sources of uncertainty [Anderson et al. 1970; Knight and Lichti 1970]. There is a 1-mg/m^3 (8-hour, time-weighted average) standard for respirable talc and mica measured with the cyclone.

The standard applied for cotton dust is not strictly for respirable dust, but, like U.S. practice, it does involve some size selection. It is 0.5 mg/m^3 (8-hour, time-weighted average) measured using a static sampler with a 2-mm-mesh grid to exclude the coarse airborne material [British Occupational Hygiene Society 1972].

"TOTAL" DUST

Various types of filter holder have been adopted in the past for sampling materials for which there is no generally accepted size-selection criterion. Most of these holders have been for 25-mm filters, because of convenience in analysis. A 25-mm open holder with a perforated cap is used by HSE for oil mist; a 47-mm holder with perforated front has been more widely used for coke oven fume [Ogden and Wood 1975]. A 25-mm holder made of metal by Casella London Ltd. with a single 4-mm orifice is used for lead [HSE 1981b], and the same sampler with seven 4-mm holes is used for wood dust, and is under consideration by HSC for use with new standards for polyvinyl chloride (PVC) dust and manmade mineral fiber.

The criterion for adoption of most of these was convenience of use with the material sampled and the analytical method used, but there has recently been growing awareness of the size selection made by all filter holders, especially at large particle sizes. The IOM in Edinburgh has taken a leading part in exploring these problems in relation to the particles actually inhaled by the exposed subject [Armbruster et al. 1983], and in investigating the performance of the personal samplers used in Britain [Wood and Birkett 1979]. The fundamental work in this field by Davies and co-workers over the past 35 years is well known. This has taken place mainly at the London School of Hygiene and Tropical Medicine, and later at the University of Essex, with the support of the Medical Research Council and other bodies.

FIBROUS DUSTS

The general appreciation of the aerodynamic properties of workplace fibrous aerosols followed from the work for the MRC by Timbrell [1965]. The membrane filter method for evaluating airborne asbestos was largely developed by the British asbestos industry [Addingley 1966], which also sponsored much of the work, through its Asbestosis Research Council, by the IOM [Beckett, 1980; Beckett and Attfield 1974] on the problems associated with this method. The HSE laboratory at Cricklewood has worked on improvement of mounting methods for membrane filters, on the associated problems of the detection limit of phase-contrast microscopy, and the preparation methods for scanning (SEM) and transmission (TEM) electron microscopy [Burdett et al. 1980]. A test slide has been developed by HSE and the British National Physical Laboratory, for standardizing the detection limit of phase-contrast microscopes in asbestos counting. This is commercially available from PTR Optics, Leeds, UK.

Following a recommendation of the Advisory Committee on Asbestos [HSC 1978], a Central Reference Scheme was established in Britain to specify more closely the details of the membrane filter method, and to provide a national quality assurance scheme for the method. New counting rules were agreed on a technical level between HSE and the asbestos industry, but have not yet been implemented because of the complicated interaction with the new control limits recommended by the advisory committee [HSC 1979a] and with similar discussion taking place in the European communities. HSE is well advanced with work to automate its asbestos counting using improved software for the Magiscan image analyzer (Joyce-Loebl Ltd.) and an automated microscope and slide changer, but full implementation of this will await agreement on counting rules.

Timbrell's [1975] work on magnetic alignment of asbestos has led to the Vickers M88 microscope (among other things), which permits asbestos samples on membrane filters to be aligned and assessed automatically by a light-scattering system.

Manmade mineral fiber was the subject of a report by a tripartite working party of the Advisory Committee on Toxic Substances, which in a majority report recommended standards of 5 fiber/ml (measured like asbestos) and 5 mg/m^3 (using the Casella seven-hole filter holder). These have yet to be agreed by the HSC [1979b].

DUST IN THE COAL-MINING INDUSTRY

Although a connection between dust and lung disease in mines was recognized in the early part of the 19th century, it was not until 1929 that British coal miners became eligible for compensation for silicosis; for a compensation claim to be successful it had to be proved that a man had worked in rock containing a high proportion of silica. It soon became clear that there was a considerable, and increasing, level of disease in South Wales mines, and in 1936 the Medical Research Council began its studies in this coalfield. The MRC reports [Bedford et al. 1943; Hart et al. 1942] showed that the disease could occur following exposure to dust containing little or no quartz. In consequence, "coalworker's pneumoconiosis" was defined as a compensable disease in its own right. It was concluded that the mass of dust breathed was likely to be the most important factor in the causation of pneumoconiosis, but, in the absence of suitable gravimetric dust sampling instruments for use in mines, attempts were made to correlate incidence of the disease with the number of particles of so-called "respirable" size (that is, those thought to be fine enough to penetrate to and be deposited in that region of the deep lung where the disease was known to be located) as collected by a TP, and counted and sized under a microscope. Following the publication of the MRC findings, some routine sampling and attempts at control of dust were initiated in the South Wales coalfield, and to a lesser extent in other British coalfields. However, it is fair to say that, at that time, the criteria for identifying the particular dust particles capable of causing the disease in question were not well understood.

The British coal industry was nationalized in 1947, and in 1949 "approved conditions" were agreed on between government and trade unions for the employment of men suffering from pneumoconiosis as established from chest radiographs. These dust standards are shown in Tables I and II and were based on TP measurements of particle number concentration. The new standards soon applied to all workplaces, although by present standards, frequency of sampling (once a year in some instances) was very sparse. The NCB established a scientific control organization, and one of the tasks of the coalfield laboratories was sampling and evaluation of airborne dust. At the outset, a considerable proportion of the sampling was carried out with paper-staining handpumps (Green and Hounam), and the results were related empirically to TP measurements.

In 1948 NCB founded the Central Research Establishment at Stoke Orchard, and part of the first research program was to examine the rationale and methodology of airborne dust sampling. The dust control research and development work was later moved to the Mining Research Establishment (MRE), Isleworth, and again, on the closure of MRE, to the present Mining Research and Development Establishment.

Walton's concept of "selective sampling of respirable dust" [Walton 1954] was, in many ways, the foundation of the work. It was based on the realization that passage through the upper respiratory passages and subsequent deposition of airborne particles in the alveoli were governed by the aerodynamic properties of the particles rather than geometric (microscope) diameter, and it was therefore logical to carry out this selection of the

Table I. "Approved Dust Conditions": Short-Period Sampling
Using the Thermal Precipitator (1949–1956)

Coal dust clouds in anthracite collieries	Not more than 650 particles/cm^3 between 1.0 and 5.0 μm in size
Coal dust clouds in other collieries	Not more than 850 particles/cm^3 between 1.0 and 5.0 μm in size
Stone drivages at collieries	Not more than 450 particles/cm^3 between 0.5 and 5.0 μm in size

Table II. Modification of the "Approved Conditions" Scheme, in Which Additonal Sampling was Carried Out When Dust Levels Approached the "Approved Levels" (1956–1965)[a]

Type of Workplace	Approved Mean Concentration (particles/cm^3)		Not Approved Mean Concentration (particles/cm^3)	
	On a Single Shift	On Each of Two Successive Shifts	On a Single Shift	On Each of Two Successive Sampling Shifts
Stone drivages[b]	≤250	250–350	>450	350–450
Cutting in dirt[c]	≤350	350–525	>650	525–650
All other locations in all coalfields except anthracite collieries in South Wales[c]	≤450	450–700	>850	700–850
All other locations in anthracite collieries in South Wales[c]	≤350	350–525	>650	525–650

[a]Places which, after two successive sampling visits, could not be categorized, remained unclassified pending further sampling visits.
[b]Particles 0.5–5 μm.
[c]Particles 1–5 μm.

respirable fraction in the sampling process. This led to the suggestion that a horizontal elutriator could be chosen for which penetration as a function of particle aerodynamic diameter would be a fair reflection of the fraction of inhaled airborne dust deposited in the alveolar parts of the lung, based on the lung deposition data available at that time. Such an elutriator therefore would have 50% penetration for particles of aerodynamic diameter 5 μ. This suggestion was approved first by the MRC Industrial Pulmonary Diseases Committee, and later by the Johannesburg Pneumoconiosis Conference [Orenstein 1960]. The wide ranges of particle size and shape related to a given falling speed were demonstrated using the "conifuge" [Hamilton 1954] and an "aerosol spectrometer" [Timbrell 1954].

The development of practical selective sampling devices incorporating horizontal elutriators, including the Hexhlet [Wright 1954] and the long-running thermal precipitator (LRTP) [Hamilton 1956] followed, and the general picture was described by Hamilton and Walton [1961].

The original dust sampling program involved roving measurements on the coal face, and obtaining a number of short-period samples. The move to a fixed sampling point came first, followed by full-shift sampling, using the LRTP at first as an alternative to the standard TP, and later in 1965 as the reference instrument. Table III shows the "approved" dust concentration levels during this period. It should be pointed out that longwall working using single roadways as intake and return had for many years been the principal technique of coal mining, so that "fixed point" or "area" sampling appeared to have obvious advantages. The MRE gravimetric dust sampling instrument [Dunmore et al. 1964] was developed, not only in response to the original suggestion that mass concentration of respirable dust was the most appropriate parameter, but also because of the difficulties of maintaining consistent results from microscopical particle sizing and counting.

Very soon after the start of the research into dust sampling, other work was initiated to look at subjects including the physics of dust formation [Hamilton and Knight 1958], the capture of airborne dust by water droplets [Walton and Woolcock 1960] and wider investigation of rock breakage [Evans and Pomeroy 1967]. By the early 1960s, the character of the coal industry was changing, from largely manual to highly mechanized. This change resulted in considerable dust problems, exacerbated by longwall mining in which, of necessity, men and machines are working in the same ventilation current. Greater mechanization did, however, have the advantage of allowing identification of dust sources. Laboratory dust suppression research [Hamilton et al. 1962] and underground trials [Evans and Hamilton 1964] laid the scientific basis for an engineering dust control program.

In 1953 NCB began a nationwide epidemiological study in 25 pits, covering the whole range of coal types, to try to establish the nature of the relationship between dust exposure and disease. It was the largest study of its type ever

Table III. "Approved Dust Conditions": Shift Length Sampling Using the LRTP (1966–1970)

Type of Workplace	Basic Standard (particles/cm^3)	Sequential System				
		Approved		Not Approved		Provisionally Approved: Where Two Successive Sampling Shifts Have Given Mean Concentration (particles/cm^3)
		Mean Concentration on a Single Shift (particles/cm^3)	Mean Concentration on Each of Two Successive Sampling Shifts (particles/cm^3)	Mean Concentration on a Single Shift (particles/cm^3)	Mean Concentration on Each of Two Successive Shifts (particles/cm^3)	
Stone drivages	250	≤130	130–200	>250	200–250	One shift, 130–200; other shift, 200–250
All other locations in all coalfields (except anthracite collieries in South Wales)	700	≤370	370–580	>700	580–700	One shift, 370–580; other shift, 580–700
All other locations in anthracite collieries in South Wales	500	≤260	260–410	>500	410–500	One shift, 260–410; other shift, 410–500

undertaken, and its ultimate aim was to allow the formulation of meaningful dust standards. It was known as the Pneumoconiosis Field Research (PFR) and in 1969 was taken over (on behalf of NCB) by IOM. Full-chest radiographs, together with occupational histories, were obtained five times per year for the whole underground workforce at the 25 pits studied. Full-shift dust sampling of different occupational groups also was carried out, using the TP. By 1963, although the research showed clear evidence of disease incidence and progression in the various pits, only a weak correlation could be demonstrated between radiological changes and dust concentration as expressed in terms of particle number. However, after the MRE gravimetric respirable dust sampler was introduced in 1964, and when all of the earlier dust exposure data were recalculated, coalfield-by-coalfield (to convert particle number concentrations to equivalent respirable mass concentrations), a greater correlation was demonstrated. These PFR measurements, together with others made with the MRE sampler in different mines [Hamilton et al. 1967], showed that there was a wide range in the relationship between the concentration measured in terms of mass respirable dust and the number of particles, over different coalfields and mining operations.

By 1969 it had become possible to calculate the long-term risk of developing the different categories of pneumoconiosis during a lifetime of work at the coal face [Jacobsen et al. 1971] (Figure 1). New dust concentration standards based on gravimetric respirable dust sampling using the MRE gravimetric sampler were introduced in all mines in 1970 [Chamberlain et al. 1971; Jacobsen et al. 1970], with the immediate aim of reducing the dust concentration below 8 mg/m^3 as measured at a prescribed fixed-point sampling position in the return roadway 70 m downwind of a longwall face. (It should be noted here that the relationships shown in Figure 3 relate to the exposure of the men on the face and not at this 70-m point.) It has been shown, however, that 8 mg/m^3 measured at this point is equivalent to about 5.5 mg/m^3 for the "portal-to-portal" exposure of an average coal-face worker. This dust standard was given legal status in 1975 and was tightened to 7 mg/m^3 in 1977 (Table IV).

NCB had taken another step in 1959, setting up a periodic X-ray scheme, with mobile units visiting every colliery in a five-year cycle; this was changed to a four-year cycle in 1976. The scheme has made possible general medical surveillance and at the same time has given valuable indication of the success of the dust control program.

The introduction of gravimetric dust sampling provided the impetus for much greater effort to reduce dust levels. Figure 2 shows the mean concentration of respirable dust on all longwall faces, which fell from 5.9 mg/m^3 in 1970 to 3.5 mg/m^3 in 1980–1981. These improvements were obtained, not only as a result of a concentrated training program and encouragement to

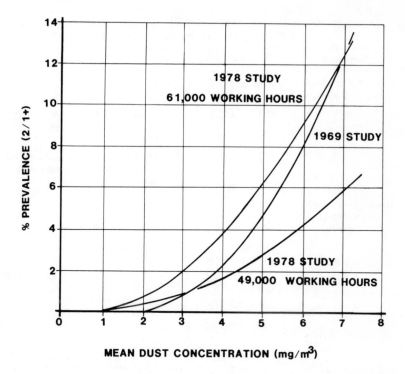

Figure 1.   Risks of a coal miner developing category 2 simple pneumoconiosis when exposed over a working lifetime to different dust concentrations on a longwall coal face. The 1969 estimates, after 10 years of research, and the 1978 estimates, after 20 years of research, are roughly similar—for 61,000 working hours. The 1978 estimates are also shown for the risks related to the reduced number of hours worked in a lifetime. (Dust concentrations measured at the coal face—"portal to portal" sampling; "% prevalence" refers to the percentage of all coal face workers contracting category 2 simple pneumoconiosis or worse during a working lifetime.)

the workforce to use and maintain dust control equipment, but by modifying mining machines to use fewer, slower-moving picks, with water sprays mounted on the cutting elements, and by improving ventilation. Exhaust ventilation and air cleaning were introduced generally in headings and recently to an increased extent on coal faces. Respirators were made generally available, as an addition to and not a substitute for dust control, and in the last year or two some use has been made of the "Airstream" ventilated helmet.

Table IV. Respirable Dust Sampling Scheme, As Amended in 1977, Using the MRE Type 113A Gravimetric Sampling Instrument[a]

Workplace	Dust Levels	Sanctions
Coalfaces: sampling 70 m in the return	≤ 7 mg/m^3	Work permitted
	>7 and ≤ 11 mg/m^3	From 5 shift measurements in two successive months: work to be stopped
	>11 mg/m^3	From 5 shift measurements followed by a further 5 shifts: work to be stopped
Headings: sampling at the position of man at greatest risk	≤ 5 mg/m^3 (if quartz conc. ≤ 0.45 mg/m^3)	Work permitted
	≤ 3 mg/m^3 (if quartz conc. >0.45 mg/m^3)	Work permitted
	>5 mg/m^3 and ≤ 7 mg/m^3 or >3 mg/m^3 (incl. quartz) ≤ 6 mg/m^3	From 5 shift measurements in two successive months: work to be stopped
	>7 mg/m^3 >6 mg/m^3 (incl. quartz)	From 5 shift measurements followed by a further 5 shifts: work to be stopped

[a]Men suffering from category 2 pneumoconiosis or younger than 35, with catetory 1, or "rapid progressors" are to work in concentrations not exceeding 2 mg/m^3.

Figures 3 and 4, which show the results of the periodic X-ray scheme (from the NCB Medical Service Report), illustrate clearly the improvement in miners' health, obtained as a result of the dust control program: the reduction in disease among the younger age groups is particularly encouraging. Notwithstanding the success achieved, some 500 men are still certified every year as having pneumoconiosis, and other respiratory complaints can be ascribed partly to dust exposure resulting in significant loss of working time.

Dust standards seem likely to be made more stringent; furthermore, increased productivity must mean greater machine utilization, so that it will be essential to improve dust control techniques. The present system of fixed-point sampling has been of value in providing a sound baseline for dust control, but it does not, of course, do more than relate to the "average" coal-face worker's exposure. Hadden et al. [1977], comparing the dust levels at different positions on longwall faces, showed that the return sampling

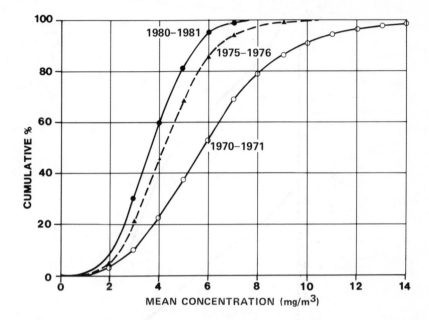

Figure 2. Cumulative percentages of UK longwall faces with mean concentration less than or equal to various levels. Shift-length sampling once or five times per month, using the MRE sampler 70 meters from the face line in the return.

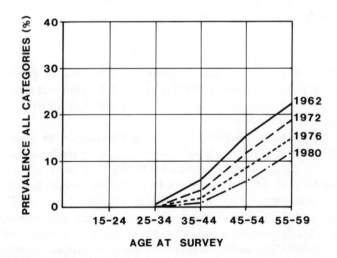

Figure 3. Prevalence of all categories of pneumoconiosis for different age groups (periodic X-ray results).

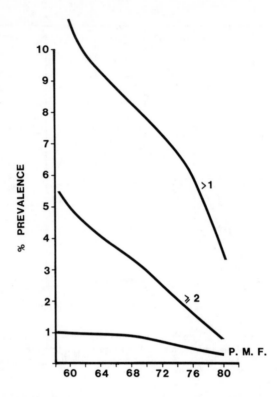

Figure 4. Prevalence of different categories of pneumoconiosis for all age groups (periodic X-ray results).

point did have much the same relationship to the bulk of face sampling positions as had been found in 1970 (i.e., about 1:4:1) but that men working at the return end of the face would have an exposure very much closer to the levels measured in the return.

In contrast to many other countries, probably because of the importance of the dust problems in the South Wales coalfield, and the results of the original MRC studies there, the role of quartz has largely been ignored in the British coal industry, and indeed the PFR results confirmed broadly that the effect of quartz was unimportant in the etiology of pneumoconiosis. However, Hurley et al. [1979] showed that there is evidence that some men may have been affected by dust with high quartz content, although quartz contents <10% in respirable dust do not appear to affect the probability of development of simple pneumoconiosis. The quartz content of respirable dust is being measured, although no control standards incorporating quartz are in force for longwall coal faces.

In addition to the MRE reference sampler, the Simslin II continuous dust monitor [Leck and Harris 1978] is used for investigational purposes [Ford et al. 1983]; Osiris [Leck 1983], a development of Simslin, appears to have some advantages as a remote transducer. A selective gravimetric sampler incorporating β-ray assessment of shift mean concentration is under development: this instrument may be used either as a self-contained unit for five-shift operation, giving a direct reading at the end of each shift, or over much longer periods to give direct readings of concentration underground and telemetered information on the surface.

DUST IN OTHER INDUSTRIES

Similar concern for the health of the workforce has been (and is being) shown by many other industries, where stringent dust standards exist and dust control measures are applied accordingly. Although it is perhaps fair to say that no other industry has received the same degree of concerted effort as the coal industry, partly because of the enormous cost involved and partly because the problem in many industries is even more difficult to study than in the coal industry, there has however been considerable progress. As described for the coal industry, the general problem in any given workplace environment comes down to deciding (1) which fraction of the airborne dust to measure in relation to which disease; (2) how to measure that fraction; (3) what relationship exists between disease and airborne levels of dust; and (4) how to reduce the level to a "safe" value.

The substantial amount of work in progress on airborne asbestos has already been mentioned: British workers, in particular those from the MRC Pneumoconiosis Unit, the IOM and the asbestos companies themselves are participating actively in the considerable international cooperative effort which is going on at present. Industries in which there are problems connected with radioactive aerosols, lead particles, some wood dusts, etc., make use of "total" dust sampling, and the various techniques employed have been discussed earlier in this chapter. There is, however, a general realization that the "total" dust is defined by the specific measurement instrument employed, and that there should be a move to measurements which represent as far as possible all the dust that would be inhaled through the nose and/or mouth during breathing.

THE FUTURE

The experience in the British coal industry has demonstrated the benefits to be gained in the work aerosol field by the balanced multidisciplinary approach. The problems to be tackled in the coal industry have been mentioned; epidemiological and animal studies are in progress at IOM to examine the problems of quartz in mixed mine dusts and the role of dust exposure in chronic bronchitis, particularly the possible contribution from particles coarser than those contained in the respirable fraction.

The concept of "inhalable" or "inspirable" dust is receiving considerable attention, both in cooperation with colleagues in Europe and the international industrial hygiene community at large. Harmonization of asbestos and manmade mineral fiber assessment is in progress.

In the great majority of workplaces other than coal mining on longwall faces, personal sampling is the norm. This raises the problem of the relationship of present sampling methods to standards derived from static samples. Many research projects in this field, too numerous to describe, are in hand; there is now a general realization that only through a proper understanding of the nature of the hazard to health, and the means by which it can be quantified, can a basis for its practical control be achieved.

REFERENCES

Addingley, C. G. (1966) "Asbestos Dust and Its Measurements," *Ann. Occup. Hyg.* 9:73-82.

Anderson, D. P., J. A. Seta and J. F. Vining (1970) "The Effect of Pulsation Dampening on the Collection Efficiency of Personal Sampler Pumps," NIOSH Report TR70.

Armbruster, L., H. Breuer, J. H. Vincent and D. Mark (1983) "Definition and Measurement of Inhalable Dust," Chapter 17, this volume.

Beckett, S. T. (1980) "The Effect of Sampling Practice on the Measured Concentration of Airborne Asbestos," *Ann. Occup. Hyg.* 23:259-272.

Beckett, S. T., and M. D. Attfield (1974) "Inter-laboratory Comparisons of the Counting of Asbestos Fibres Sampled on Membrane Filters," *Ann. Occup. Hyg.* 17:85-96.

Bedford, T., et al. (1943) "Chronic Pulmonary Diseases in South Wales Coalmines II Environmental Studies," Medical Research Council Special Report Series No. 244, His Majesty's Stationery Office, London.

Blackford, D. B., and G. W. Harris (1978) "Field Experience with Simslin II– A Continuously Recording Dust Sampling Instrument," *Ann. Occup. Hyg.* 21:301-313.

British Occupational Hygiene Society (1972) "Hygiene Standards for Cotton Dust," *Ann. Occup. Hyg.* 15:165-192.

Burdett, G., J. M. M. le Guen, A. P. Rood and S. J. Rooker (1980) "Comprehensive Methods for Rapid Quantitative Analysis of Airborne Particulates by Optical Microscopy, SEM and TEM with Special Reference to Asbestos," in *Atmospheric Pollution 1980*, M. M. Benarie, Ed. (Amsterdam: Elsevier).

Chamberlain, E. A. C., A. D. Makower and W. H. Walton (1971) "New Gravimetric Dust Standards and Sampling Procedures for British Coal Mines," in *Inhaled Particles III*, W. H. Walton, Ed. (Old Woking, UK: Unwin Bros.).

Dunmore, J. H., R. J. Hamilton and D. S. G. Smith (1964) "An Instrument for the Sampling of Respirable Dust for Subsequent Gravimetric Assessment," *J. Sci. Instr.* 41:669-672.

Evans, C. G., and R. J. Hamilton (1964) "Production of Dust by Anderton Shearers," *Colliery Guardian* 209:445-454.

Evans, I., and C. D. Pomeroy (1967) *Strength Fracture and Workability of Coal* (London: Pergamon Press, Ltd.).

Ford, V. H. W., R. Minton and D. Mark (1982) "Comparative Trials in Coal Mines of the TM Digital and Simslin Dust Monitors," in *Aerosols in the Mining and Industrial Work Environment, Vol. 3, Instrumentation,* V. A. Marple and B. Y. H. Liu, Eds. (Ann Arbor, MI: Ann Arbor Science Publishers).

Hamilton, R. J. (1954) "The Relation Between Free-Falling Speed and Particle Size of Airborne Dusts," *Brit. J. Appl. Phys., Suppl. 3.*

Hadden, G. G., C. O. Jones and H. L. Thorpe (1977) "A Comparative Assessment of Dust Surveillance Procedures, Including the Use of 'Personal' and 'Fixed Position' Sampling Instruments," Final Report on CEC Contract 6253-21/8/015, Institute of Occupational Medicine Report No. TM/77/15 (EUR P41).

Hamilton, R. J. (1956) "A Portable Instrument for Respirable Dust Sampling," *J. Sci. Instr.* 33:87-108.

Hamilton, R. J. and G. Knight (1958) "Some Studies of Dust Size Distribution and the Relationships Between Dust Formation and Coal Strength," in *Mechanical Properties of Non-metallic Brittle Materials,* W. H. Walton, Ed. (London: Butterworth Science Publishers).

Hamilton, R. J. and W. H. Walton (1961) "The Selective Sampling of Respirable Dust," in *Inhaled Particles and Vapours,* C. N. Davies, Ed. (Oxford: Pergamon Press, Ltd.).

Hamilton, R. J., M. L. Levin and K. W. McKinlay (1962) "Research into the Formation and Suppression of Dust at Fast Moving Cutter Picks," *Min. Eng.* 121 (21).

Hamilton, R. J., G. D. Morgan and W. H. Walton (1967) "The Relationship Between Measurement of Respirable Dust by Mass and by Number in British Coal Mines," in *Inhaled Particles II,* C. N. Davies, Ed. (London: Pergamon Press, Ltd.).

Hart, P. A. and E. A. Aslett (1942) "Chronic Pulmonary Diseases in South Wales Coalmines I Medical Studies," Medical Research Council Special Report Series No. 243, His Majesty's Stationery Office, London.

Health and Safety Commission (1978) "Asbestos—Measurement and Monitoring of Asbestos in Air, 2nd Report by the Advisory Committee on Asbestos," Her Majesty's Stationery Office, London.

Health and Safety Commission (1979a) "Asbestos. Volume 1: Final Report of the Advisory Committee," Her Majesty's Stationery Office, London.

Health and Safety Commission (1979b) "Man-made Mineral Fibres. Report of a Working Party to the Advisory Committee on Toxic Substances," Her Majesty's Stationery Office, London.

Health and Safety Executive (1978) "Toxic Substances: A Precautionary Policy," Guidance Note EH18, Her Majesty's Stationery Office, London.

Health and Safety Executive (1981a) "Threshold Limit Values 1980," Guidance Note EH15/80, Her Majesty's Stationery Office, London.

Health and Safety Executive (1981b) "Control of Lead: Air Sampling Techniques and Strategies," Guidance Note EH28, Her Majesty's Stationery Office, London.

Higgins, R. I. and R. Dewell (1966) "A Gravimetric Size-Selecting Personal Dust Sampler," in *Inhaled Particles and Vapours II,* C. N. Davies, Ed. (Oxford: Pergamon Press, Ltd.).

Hurley, J. F., L. Copland, J. Dodgson and M. Jacobsen (1979) "Simple Pneumoconiosis and Exposure to Respirable Dust: Relationships from Twenty-Five Years' Research at Ten British Coal Mines," Institute of Occupational Medicine Report No. TM/79/13.

Jacobsen, M., S. Rae, W. H. Walton and J. M. Rogan (1970) "New Dust Standards for British Coal Mines," *Nature* 277:445.

Jacobsen, M., S. Rae, W. H. Walton and J. M. Rogan (1971) "The Relation Between Pneumoconiosis and Dust Exposure in British Coal Mines," in *Inhaled Particles III,* W. H. Walton, Ed. (Old Woking, UK: Unwin Bros.).

Knight, G. and K. Lichti (1970) "Comparison of Cyclone and Horizontal Elutriator Size Selectors," *Am. Ind. Hyg. Assoc. J.* 31:437-441.

Leck, M. J. (1983) "Optical Scattering Instantaneous Respirable Dust Indication System," in *Aerosols in the Mining and Industrial Work Environment, Vol. 3, Instrumentation,* V. A. Marple and B. Y. H. Liu, Eds. (Ann Arbor, MI: Ann Arbor Science Publishers).

Leck, M. J. and G. W. Harris (1978) "Simslin II," *Colliery Guardian* 226: 676-677.

Maguire, B. A., D. Barker and D. Wake (1973) "Size-Selection Characteristics of the Cyclone Used in the Simpeds 70 Mk 2 Gravimetric Dust Sampler," *Staub* 33:95-99.

Ogden, T. L. and J. D. Wood (1975) "Effects of Wind on the Dust and Benzene-Soluble Matter Captured by a Small Sampler," *Ann. Occup. Hyg.* 17:187-196.

Orenstein, A. J. (1960) "Recommendations Adopted by the Pneumoconiosis Conference," in *Proceedings of the Pneumoconiosis Conference, Johannesburg, 1959,* A. J. Orenstein, Ed. (London: Churchill) pp. 619-621.

Timbrell, V. (1954) "The Terminal Velocity and Size of Airborne Dust Particles," *Brit. J. Appl. Phys.,* Suppl. 3.

Timbrell, V. (1965) "The Inhalation of Fibrous Dusts," *Ann. N.Y. Acad. Sci.* 132:255-273.

Timbrell, V. (1975) "Alignment of Respirable Fibres by Magnetic Fields," *Ann. Occup. Hyg.* 18:299-312.

Walton, W. H. (1954) "Theory of Size Classification of Airborne Dust Clouds by Elutriation," *Brit. J. Appl. Phys.* Suppl. 3.

Walton, W. H., and A. Woolcock (1960) "The Suppression of Airborne Dust by Water Spray," in *Aerodynamic Capture of Particles,* E. G. Richardson, Ed. (Oxford: Pergamon Press, Ltd.).

Wood, J. D. and J. L. Birkett (1979) "External Airflow Effects on Personal Sampling," *Ann. Occup. Hyg.* 22:299-310.

Wright, B. M. (1954) "A Size-Selecting Sampler for Airborne Dust," *Brit. J. Ind. Med.* 11:284-288.

CHAPTER 2

STATUS OF WORK-ENVIRONMENT AEROSOLS IN THE FEDERAL REPUBLIC OF GERMANY

W. Koch and W. Stöber

Fraunhofer-Institute for Toxicology and Aerosol Research
Hannover Münster, Federal Republic of Germany

ABSTRACT

This chapter describes the status of work-environment aerosols in the Federal Republic of Germany. The complicated system of basic laws, ordinances and technical guidelines to regulate the abatement and control of work-environment aerosols is presented. The instruments for measuring the relevant workplace aerosol concentrations are described in detail. We conclude with a brief discussion of the statistical evaluation procedures necessary for the assessment of the workplace dust situation.

INTRODUCTION

The Federal Republic of Germany is a highly industrialized country, and a substantial part of the West German industrial workforce labors in plants, factories and mines where workers may have contact with industrial products and wastes containing or consisting of substances that may be dangerous to human health. One of the oldest and most serious examples is the exposure of coal miners to coal dust as a by-product of the mining process, which may cause pneumoconiosis, a serious lung impairment. The importance of noxious dusts as a cause of occupational diseases is shown in Table I. These figures are based on statistical data collected by the German industrial vocational associations [Heidermanns et al. 1980]. The fraction of occupational

21

Table I. Occupational Diseases in West Germany (1978)

Occupational Diseases	Registered		Compensated		Fatal		
	N_r	%	N_c	%	N_f	% (N_c)	% (N_f)
Total	41,470	100	6582	100	148	2.2	100
Caused by Dusts	6,100	15	1365	21	128	9.4	86

diseases caused by dust grows from about 15% of the registered cases up to 86% for the cases which have caused the death of the victim.

Soon after the beginning of the industrialization in the 19th century, a system of regulations and laws started to evolve in Germany that laid down the rules by which the workers were to be protected against dangerous conditions and noxious effects in the workplace. Nowadays, this complex system of rules, along with other legislation, regulates promulgation of threshold limit values (TLV) of noxious dusts, mandatory preventive control technology for reducing dust formation, limitation of working hours in dusty environments, compulsory medical checkups of the workers and monitoring work-environment dust.

One purpose of this chapter is to excerpt and summarize from today's West German laws on workplace safety the regulations that are the legal base for abatement and control of work-environment aerosols. The other purpose is to describe the present status of instrumentation and measuring techniques for dusts and aerosols that has developed from that safety legislation.

GENERAL GOVERNMENTAL REGULATIONS

The German safety regulations have a three-pronged legal foundation on which detailed rules and ordinances are based:

1. industrial code and trade regulations (Gewerbeordnung);
2. mining code (Bergordnung); and
3. insurance code (Reichsversicherungsordnung).

The first and second codes cover all typical aspects of industrial, commercial and mining operations. They contain, among other rules, general statements on the physical and social protection of workers in the various industries. On the basis of these general laws, the Ministry of Labor has developed and promulgated federal ordinances regulating in detail workplace conditions for specific operations or precautions to be taken in handling and controlling certain substances in working places ["Verordnung über Gefähr-liche Arbeitsstoffe" 1975]. Compliance with the rules and regulations is supervised in general by the trade inspection offices (Gewerbeaufsichtsämter) and, in particular for the mining industry, by the mining offices (Bergämter).

The third code regulates basically the responsibilities for the physical and social welfare of the working population. In particular, this law is the legal base of various compulsory and trade-specific social security insurances as well as mandatory health and accident insurance, the latter protecting business against liability and compensation claims from work accidents. Accident insurance is established and operated by vocational associations, and the law authorizes these associations to issue binding safety rules. One of these rules is the ordinance VBG 119 [1973] on the "protection against health-endangering mineral dust," a typical rule to control some noxious substances. This rule is to be viewed in context with the regulations about the corresponding working places.

MAXIMALE ARBEITSPLATZ-KONZENTRATIONEN (MAK) AND TECHNISCHE RICHT-KONZENTRATIONEN (TRK) LISTS, TECHNICAL RULES AND GUIDELINES

To avoid frequent modifications of the basic laws, these laws are phrased in rather general terms and do not account for the actual state-of-the-art regarding control technology and instrumentation or the actual value of the threshold limit. Instead, the basic laws are complemented by lists of binding threshold limit concentrations, technical rules and guidelines which are established by expert panels. A list of the threshold values is the list of the "maximum workplace concentrations" (Maximale Arbeitsplatz-Konzentrationen) [Senatskommission zur Prüfung Gesundheitsschädlicher Arbeitsstoffe 1979]. It contains threshold limit values in milligrams per cubic meter for single components in the gaseous and/or particulate airborne state. These values are established on the basis of toxicological or epidemiological evidence. Usually, except for fibrogenic dust, they are mean values over a whole working day. Limit values for fibrogenic dusts are long-term mean values and relate to a dust exposure of one year or, if the individual dust exposure is documented, they relate to a dust exposure of five years.

In addition to the limiting values, the MAK list also gives an aerosol definition that is supposed to be related to potential health effects. According to this definition, the characteristic size parameter for particles and fibers is the aerodynamic diameter, which essentially governs the aerodynamic behavior of the particles and, thus, the mechanism of the particle deposition in the lung. Then, the conventional definition of the inhalable fraction of the dust is given by the fraction of the total dust that is collected by an instrument drawing the air into its sampling head at a flowrate of 1.25 m/sec (±10%). The fraction of airborne matter that may penetrate deeply into the lung and reach the alveolar region is called fine dust. This fraction is defined as passing through a horizontal sedimentation precipitator (elutriator) designed and operated to retain 50% of the particles with an aerodynamic diameter of 5 μm. The actual retention curve is shown in Figure 1 (Johannesburg Convention).

For health-endangering substances that for medical or other reasons, do not permit the assignment of a safe MAK value, "Technical Guiding Concentrations" (Technische Richt-Konzentrationen) are established annually [TRgA 102 1979]. This is the case, for instance, for carcinogenic substances. In contrast to the MAK values, the TRK values do not necessarily indicate that the health risk for the worker being exposed to these substances is reduced to a reasonably low level. They are mainly determined by their technical feasibility and set standards according to the best technology available. They will be reduced whenever technically possible.

There is an additional, generally accepted dust classification scheme that relates to various specific health-endangering effects [Heidermanns et al. 1980]. This scheme is shown in Table II. Whether MAK- or TRK-values are valid will depend on the noxious substances associated with the particles.

On the basis of the requirements set forth by the MAK and TRK lists, detailed technical rules and guidelines have been established by expert panels. These documents contain precise instructions about the measuring strategies to be pursued, the types of instruments to be used and the way the results of

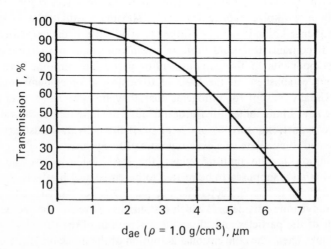

Figure 1.   Transmission curve according to the Johannesburg convention.

Table II. Dust Categories by Medical Indications

Dust Categories	Examples with MAK or TRK Values (mg/m^3)	To Be Measured
Fibrogenic	Quartz, MAK 4/0.15[a]	Fine dust
Toxic	Lead, MAK 0.1	Total dust
Radioactive	Uranium, MAK 0.25	Total dust
Allergenic	Mercury compounds, MAK 0.01	Total dust
Carcinogenic	Asbestos, TRK 2/0.05[b]	Total dust, fine dust, fibers
Inert	Titanium dioxide, MAK 8	Fine dust

[a]First/second number for a weight percentage of quartz smaller/greater than 3.75.
[b]First/second number for a weight percentage of asbestos smaller/greater than 2.5.

the measurements should be interpreted. They do not have the force of laws and ordinances, but are technical guidelines for the surveillance institutions and the courts. Such rules and guidelines are:

- "Technical Rules for Dangerous Substances in Workplaces," (Technische Regeln für Gefährliche Arbeitsstoffe) created by the committee for dangerous substances in the workplace [TRgA 401 1979];
- "Rules for the Measurement and Interpretation of Health-Endangering Dusts of the VBG" [Hauptverband der Gewerbliche Berufsgenossenschaften 1979; and
- Verein Deutscher Ingenieure (VDI) guideline No. 2265 "Determination of Dust Concentrations in Workplaces for Industrial Hygiene Purposes" [VDI-Richtlinie No. 2265 1980].

With regard to measuring techniques and instrumentation, all of the three guidelines are practically identical. Essentially, a standard procedure for measuring aerosols in workplaces is recommended meeting the following requirements:

1. A fine dust fraction will be measured in accordance with the Johannesburg Convention.
2. The sample will be taken at a level above ground corresponding to the level of the head of a man in a typical working position. The sampling locations will be representative for the particular work process. The sampled volume of air should not exceed 10% of the volume of the room.
3. The sampling periods will be distributed in time over a normal work shift, and the total sampling time will not be less than one hour. If instruments with smaller sampling periods are used, samples should be spaced equally in time and total at least one hour.
4. Chemical analysis of the aerosol sample will comply with the specifications of the respirable dust as defined for the various TLV.

In certain well defined cases, the actual measuring procedure may be different from the standard procedure as long as conversion factors and functions can be applied that do not exceed corrections by a factor between 0.7 and 1.5. The geometric standard deviation should be less than 1.5. The measuring instruments should measure values in the physical units according to the MAK definition. Another exception from standard measurements may be permitted for the class of the indirect measuring procedures that allow a conversion to standard procedures only if the physical state of the aerosol does not change during measurement. Generally, these procedures are less suitable for assessment of workplace dust.

INSTRUMENTATION

Most of the sampling instruments involve a gravimetric dust measurement that is related directly to the MAK and TRK definitions. In contrast to some optical devices, gravimetric instruments avoid an overestimate of the extremely small particles that will not deposit in the alveolar range of the lung but can occasionally be found in considerable number concentrations in workplaces, for instance, originating from the exhaust of diesel engines in salt mines. Therefore, present routine and surveillance measurements in working places in West Germany are almost exclusively gravimetric measurements. An exception is the determination of asbestos fibers: the number concentration may be determined microscopically because TRK values are alternatively given in equivalent numbers of particles per cubic meter.

In sampling the total amount of dust, airborne particulate matter is drawn into the sampling head and the particles may simply be deposited on a suitable membrane or fiber filter. A table showing the relevant physical characteristics of some filters is given in VDI-Richtlinie 2265 [1980]. The upper and lower limit of the size range that can be registered depends on the efficiency curve of the instrument and the filter.

A typical commercial air sampler for application in workplace areas is the VC 25 G instrument (available from Sartorius GmbH, Göttingen, West Germany [Coenen 1975]). This stationary sampler uses a blower and a flow-controlled power supply to maintain a constant air flowrate at 22.5 m^3/hr independent of the pressure drop at the filter. The sampling head is a circular slit; the sampling velocity at the intake can be varied between 1 and 2 m/sec.

For fine dust measurements it is advantageous to effect size separation during sampling rather than in a subsequent batch analysis employing, for instance, a sedimentation chamber for the resuspended dust. Therefore, various instruments have been developed using precipitators or elutriators to separate the coarse dust from the respirable fraction. Three stationary samplers will be described below in detail, as they use different precipitating mechanisms, i.e., sedimentation, centrifugal deposition and inertial impaction of particles. All of these mechanisms are typical for separating aerosol particles with respect to their aerodynamic diameter.

An instrument using a horizontal elutriator is the MPG II (Fa. Wazau, Berlin) [Bauer and Bruckmann 1974]. Figure 2 shows a longitudinal cross section of the instrument. A laminar flow of air is drawn through a system of parallel plates. A critical orifice provides a constant flow independent of both the pressure drop across the filter and the pressure fluctuation of the suction device. This can be either a vacuum pump or a pneumatic ejector. The pressure drop across the filter, which is a measure of the mass of dust deposited on the filter, is displayed and can be registered externally. The geometry of the horizontal elutriator is adjusted in such a way that the final separation curve will match the curve of the Johannesburg Convention. In this case, the value of the conversion factor is one and, thus, the MPG II may serve as a standard instrument complying with the MAK definition. Deviations from an ideal performance, as they are possibly encountered in practice, are indicated by the shaded areas in Figure 3.

Another frequently employed instrument is the TBF 50 [Breuer 1971; Stuke and Emmerichs 1973] (Figure 4), manufactured by Mollidor und Müller, Rodenkirchen, FRG. This device samples at a flowrate of 50 liter/min and consists of two cyclones, the first one precipitating the coarse particles and the second one sampling the fine dust fraction. As can be seen from Figure 2, ultrafine particles that are not retained in the lung pass through the cyclones and are possibly deposited on a filter. The shape of the efficiency curve of the second cyclone causes the conversion factor to depend on the size distribution of the aerosol to be sampled. (The conversion factor relates the concentration measured with the TBF 50 to the concentration available from MPG II measurements.) This effect has been measured by Reisner [1975]. With a small amount of dust, the TBF 50 overestimates the presence of fine dust in an aerosol (conversion factor <1), but the conversion factor is >1 if the relative amount of fine dust is high. In all cases, however, the conversion factor is within the limits established in the technical guidelines.

A measuring system employed by the technical surveillance services of the vocational associations is the VC 25 F instrument (flowrate 22.5 m^3/hr) [Coenen 1975]. Here the larger particles are precipitated by inertial forces. For this purpose the sampling head of the VC 25 G is replaced by the device shown in Figure 5. Apparently, the design resembles the stage of an impactor. Small particles follow the streamlines and are deposited uniformly over the filter. Large particles with a sufficiently adhesive surface will impact in a small circular region below the inlet tube, whereas large particles with insufficient adhesion will be reflected and sampled in the peripheral region of the filter. The annular intermediate region is free from coarse dust [Zebel 1973]. The collection efficiency curve is shown in Figure 3.

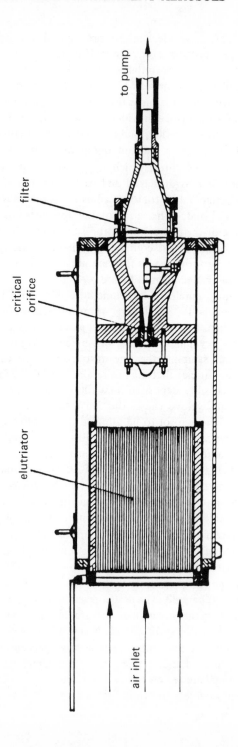

Figure 2.   The MPG II device.

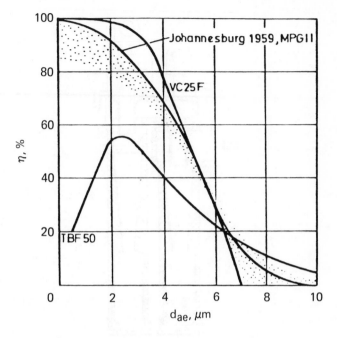

Figure 3.   Collection efficiency of various instruments.

The three instruments just described are examples of designs with a record of satisfactory practical performance and simple operating requirements. In general, all precipitating systems retaining 50% of particles with an aerodynamic diameter of 5 μm are applicable for fine dust sampling if their size variation coefficient is not more than 0.25 [Schütz and Coenen 1974]. For several instruments, Schütz and Coenen calculated mean conversion factors depending on the respective variation coefficient. In their calculations, they used two extreme size distributions and obtained a mean conversion factor as the geometric mean of the conversion factors related to the two extreme distributions. Most practical dusts fall somewhere in between the extreme cases.

In most cases, simple gravimetric devices are not suited to give instantaneous results. However, if, for instance, it is necessary to assess the actual fluctuations of the dust concentration in time as dependent on certain operational procedures in a workplace, on-line measurements are practically indispensable. For such purpose, optical instruments are of great advantage. An instrument of this kind is the "Tyndallometer digital" [Breuer and Robock 1975], an optical photometer that is an improved version of the old Tyndalloscope widely used in the 1950s for routine measurements in the mining industry

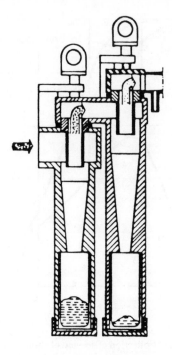

Figure 4. The TBF 50 device.

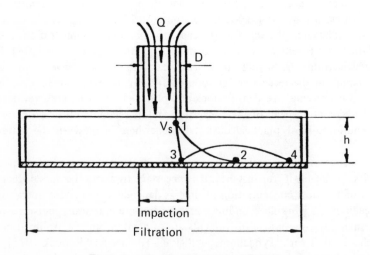

Figure 5. Sampling head of the VC 25 F.

(Figure 6). In this "semistationary" instrument, infrared light ($\lambda = 0.94\ \mu m$) emitted from a periodically interrupted light source is scattered by the dust particles. The light scattered in the direction of $70°$ relative to the direction of incidence is detected by a photodiode. Its intensity is proportional to the volume concentration of the particles inside the measuring chamber. Additional electronic devices prevent the results of the measurements from being influenced by external light. Therefore, the measuring chamber can be open. The instrument posesses analog and digital signal outputs. The optical parameters are chosen so that the assessment curve of the instrument is similar to the separation curve of the gravimetric dust samplers (Figure 7). Armbruster et al. [1977] showed the linear correlation between concentration values measured with the TM digital and those measured with various gravimetric instruments (Figure 8).

Besides the stationary instruments, there are various personal dust samplers for fine dust measurements, for example the Simpeds 70, developed in England. The Research Institute for Silicosis has tested this instrument and consented that it is suitable for practical purposes. We shall not go into further details here.

Based on the technical guidelines, sophisticated but economic routines have been developed for mass determination of the sampled dust by weighing or β-ray extinction and for separate determinations, for instance, of the quartz fractions by X-ray scattering. This facilitates rapid analysis of the samples [Coenen 1975; Emmerichs et al. 1975,1977]. For common chemical analysis routines, further detailed instructions are given in the guideline on analysis of the air produced by a task group on analytical chemistry of the Deutsche Forschungsgemeinschaft [Henschler 1972]. These guidelines will be sufficient

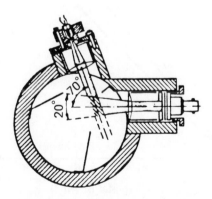

Figure 6.   Optics of the TM digital.

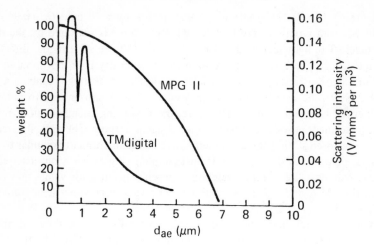

Figure 7.   Assessment curve of the TM digital.

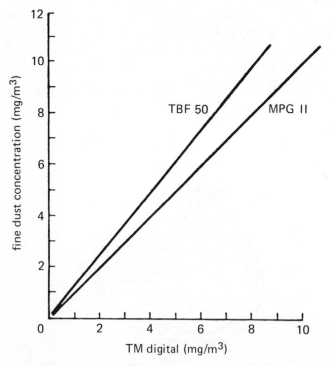

Figure 8.   Correlation between the TM digital and the gravimetric dust samplers TBF 50 and MPG II.

for most practical applications. No detailed discussion of the various aspects of chemical analyses of airborne particulate matter will be attempted here; it would exceed the scope of this chapter.

STATISTICS

Besides standard sampling and chemical analysis, a uniform statistical evaluation of the results is another important requirement to facilitate comparison of statements about the dust burden in various workplaces. Standard statistical data reduction on the basis of the actual measurements will then permit an objective and intercomparable assessment as to whether or not a MAK of TRK value G is or will be exceeded. The basic rules of these mathematical operations are the background of the detailed instructions on statistical evaluations laid down in the technical guidelines. They are sufficient for most practical applications.

There are two different statistical evaluation procedures. One is laid down in VDI guideline 2265 [1980] and VBG 119 [1973]; the other evaluation is given in TRgA 401 [1979]. The first procedure is used for the assessment of asbestos and dusts for which MAK values are given; the second procedure deals with dusts to which TRK values are assigned. No details will be reported; instead, a rough outline of the first evaluation procedure is given. For prospective estimates, which, for instance, cover an assessment period of one year, two functions $f_1(t_e)$ and $f_2(t_e)$, where t_e is the effective observation time, are tabulated for practical purposes (Table III). These functions are established on the basis of a large amount of empirical data and apply to a normal period of continuous dust exposure (8 hours per day, max. 45 hours per week). Within the range of $f_1^{-1}(t_e) \geq G_m/G \geq f_2^{-1}(t_e)$, where G_m is the geometric mean of the measured concentrations, no decision is possible. If G_m/G is larger than $f_1^{-1}(t_e)$ [smaller than $f_2^{-1}(t_e)$], the TLV is [is not] exceeded. Deviations from normal exposure periods can be taken into account by correction factors. More detailed information about the statistics can be taken from the guidelines or the basic mathematical literature [Coenen 1971,1976; Riediger and Coenen 1975].

REFERENCES

Armbruster, L., H. Breuer and G. Neulinger (1977) "Weiterentwicklung und Erprobung des Feinstaub-Streulichtfotometers," in *Ergebnisse von Untersuchungen auf dem Gebiet der Staub- und Silikosebekämpfung im Steinkohlenbergbau Vol. 11* (Essen, FRG: Verlag Glückauf), p. 107.

Table III. Assessment Factors

t_e (hr)	$f_1(t_e)$	$f_2(t_3)$
1	0.38	4.10
2	0.40	3.92
3	0.41	3.81
4	0.42	3.73
6	0.43	3.62
10	0.45	3.49
20	0.47	3.31
35	0.50	3.16
60	0.52	3.02
100	0.54	2.88
200	0.58	2.69
350	0.62	2.54
600	0.66	2.38
1000	0.70	2.24
2000	0.77	2.02
3500	0.86	1.82
6000	0.98	1.59
8760	1.25	1.25

Bauer, H. D., and E. Bruckmann (1974) "Neues Gravimetrisches Staubsammelgerät zur Messung des Atembaren Feinstaubes nach den MAK-Werten," *Kompass* 84:4.

Breuer, H. (1971) "Das Gravimetrische Staubprobenahmegerät TBF 50 zur Feststellung der Konzentration, der Zusammensetzung und der Feinheit des Grob- und Feinstaubes," in *Ergebnisse von Untersuchungen auf dem Gebiet der Staub- und Silikosebekämpfung im Steinkohlenbergbau Vol. 8* (Essen, FRG: Verlag Glückauf), p. 47.

Breuer, H., and K. Robock (1975) "Das Tyndallometer TM Digital zur Unmittelbaren Bestimmung der Feinstaubkonzentrationen in Ergänzung zu Langzeitwerten Gravimetrischer Staubmeßgeräte," in *Ergebnisse von Untersuchungen auf dem Gebiet der Staub- und Silikosebekämpfung im Steinkohlenbergbau Vol. 10* (Essen, FRG: Verlag Glückauf), p. 77.

Coenen, W. (1971) "Messung und Beurteilung der Konzentration Gesundheitsschädlicher, Insbesondere Silikogener Stäube an Arbeitsplätzen der Obertägigen Industrie," *Staub-Reinhalt. Luft* 31:484.

Coenen, W. (1975) "Feinstaubmessung mit dem VC 25. Neuere Untersuchungen und Praktische Erfahrungen," *Staub-Reinhalt. Luft* 35:12.

Coenen, W. (1976) "Beschreibung des Zeitlichen Verhaltens von Schadstoffkonzentrationen durch einen Stetigen Markowprozeß," *Staub-Reinhalt. Luft* 36:240.

Emmerichs, M., R.-W. Schliephake and J. Stuke (1975) "Die Zentrale Auswertung von Feinstaubproben Betrieblicher Gravimetrischer Staubmessungen," in *Ergebnisse von Untersuchungen auf dem Gebiet der Staub-*

und Silikosebekämpfung im Steinkohlenbergbau Vol. 10 (Essen, FRG: Verlag Glückauf), p. 101.

Emmerichs, M., R.-W. Schliephake and J. Stuke (1977) "Weiterentwicklung und Ergebnisse der Zentralen Auswertung Betrieblicher Feinstaubmessungen," in *Ergebnisse von Untersuchungen auf dem Gebiet der Staub- und Silikosebekämpfung im Steinkohlenbergbau Vol. 11* (Essen, FRG: Verlag Glückauf, p. 99.

Hauptverband der Gewerblichen Berufsgenossenschaften (1979) *Regeln zur Messung und Beurteilung Gesundheitsgefährlicher Mineralischer Stäube* (ZH 1/561) (Köln, FRG: Carl Heymanns Verlag).

Heidermanns, G., G. Kuhnen and G. Riediger (1980) "Messung und Beurteilung Gesundheitsgefährlicher Stäube am Arbeitsplatz," *Staub-Reinhalt. Luft* 40:9.

Henschler, D. (1972) *Luftanalysen* (Weinheim, FRG: Verlag Chemie).

Reisner, M. T. R. (1975) "Messung des Feinstaubes mit Verschiedenen Gravimetrischen Probenahmegeräten im Steinkohlenbergbau," in *Ergebnisse von Untersuchungen auf dem Gebiet der Staub- und Silikosebekämpfung im Steinkohlenbergbau Vol. 10* (Essen, FRG: Verlag Glückauf), p. 89.

Riediger, G., and W. Coenen (1975) "Meßplanung und Statistische Beurteilung (Voraussetzungen, Wirksamkeit und Grenzen eines Modells)," *Staub-Reinhalt. Luft* 35:445.

Schütz, A., and W. Coenen (1974) "Feinstaub: Definition-Meßverfahren," *Staub-Reinhalt. Luft* 34:323.

Senatskommission zur Prüfung Gesundheitsschädlicher Arbeitsstoffe, Ed. (1979) *Maximale Arbeitsplatzkonzentration* Deutsche Forschungsgemeinschaft, Bonn–Bad Godesberg, FRG.

Stuke, J., and M. Emmerichs (1973) "Das Gravimetrische Staubprobenahmegerät TBF 50," in *Ergebnisse von Untersuchungen auf dem Gebiet der Staub- und Silikosebekämpfung im Steinkohlenbergbau Vol. 9* (Essen, FRG: Verlag Glückauf), p. 47.

TRgA 102 (1979) "Technische Richtkonzentrationen für Gefährliche Arbeitsstoffe," in *Merkblätter für Gefährliche Arbeitsstoffe* (Munich, FRG: Verlag Moderne Industrie).

TRgA 401 (1979) "Messung und Beurteilung von Konzentrationen Giftiger und Gesundheitsschädlicher Arbeitsstoffe in der Luft. Bl. 1: Anwendung von Technischen Richtkonzentrationen," in *Merkblätter für Gefährliche Arbeitsstoffe* (Munich, FRG: Verlag Moderne Industrie).

"Verordnung über Gefährliche Arbeitsstoffe," (1975) in *Merkblätter für Gefährliche Arbeitsstoffe* (Munich, FRG: Verlag Moderne Industrie).

VBG 119 (1973) "Unfallverhütungsvorschrift. Schutz gegen Gesundheitsgefährlichen Mineralischen Staub," (Köln, FRG: Verlag Carl Heymanns).

VDI-Richtlinie No. 2265 (1980) *Feststellen der Staubsituation am Arbeitsplatz zur Gewerbehygienischen Beurteilung* (Düsseldorf, FRG: VDI-Verlag).

Zebel, G. (1973) "Theoretische Betrachtungen zur Trennung der Korngrößen in einem Trägheitsvorabscheider unter Ausnutzung der Reflexion der Teilchen," *Staub-Reinhalt. Luft* 33:104.

CHAPTER 3

SAMPLING AEROSOLS IN SOUTH AFRICAN MINES

M. J. Martinson
Department of Mining Engineering
University of the Witwatersrand
Johannesburg, South Africa

ABSTRACT

Work-related respiratory diseases were first recognized as a major health hazard in Witwatersrand gold mines at the turn of the century, and early responses to the hazard are reviewed in some detail because of their influence on current practice in Witwatersrand mines and developments in other sectors of the minerals industry. The first ad hoc dust measurements in Witwatersrand mines were made in 1902 in connection with an official inquiry into silicosis on the Witwatersrand, but routine sampling was introduced in 1914 in response to new health and safety regulations promulgated in 1911; this innovation undoubtedly played a major role in bringing the hazard under control. In later years similar regulatory measures were applied to all mines nationwide, and current regulations relating to air quality and sampling are examined in some detail before reviewing five instruments used in South African mines to comply with statutory provisions and for other purposes. The importance of establishing a valid rationale for any sampling program is stressed under sampling strategy and is related to five different strategies employed for different purposes in South African mines. The chapter concludes with some general observations on the sampling of aerosols in mines.

INTRODUCTION

The large and burgeoning minerals industry in South Africa has its origins in the discovery of diamonds at Kimberley in 1869 and gold on the Witwatersrand in 1886. Since then, both sectors have survived political turmoils and fluctuating economic conditions, but, for much of the period, gold mining has particularly served as a cornerstone of the national economy. Since World War II, exploitation of other minerals— notably platinum, uranium, coal, asbestos, copper, manganese and iron ore—has increased dramatically; nevertheless, in terms of size and economic importance, gold mining continues to dominate the South African minerals industry. The figures quoted in Table I, abstracted from *Mining Statistics 1979* [1980], may help to place the various sectors of the industry in perspective.

Table I. Mineral Production, Sales Value and Number of Persons Employed in the South African Minerals Industry, 1979

| | | | Average Number of Persons (1000s) | | |
| | | | | At Work During the Main Shift | |
Sector	Production (tons)	Sales Value (10^6 R[a])	In Service	Surface	Underground
Gold	703.5	5,848	455.6	99.5	333.2
Silver	100.7	30	b	b	b
Uranium	5,637	c	b	b	b
Diamonds	8,400,000 carats	547	21.7	15.5	5.1
Coal[d]	103,800,000	1,143	120.5	65.2	50.9
Copper[e]	190,600	249	14.8	8.4	5.8
Tin[e]	2,697	29	3.4	1.5	1.7
Asbestos	249,000	107	12.1	7.4	4.7
Chrome[f]	3,297,000	89	11.5	4.3	7.1
Manganese[g]	5,182,000	175	8.3	4.5	3.5
Iron Ore[h]	31,600,000	294	8.8	6.0	2.5
Quarries and Salt	i	i	17.3	9.1	8.2
Other	j	j	45.6	14.1	30.6
Works	k	k	83.8	83.8	

[a]R = rand, the monetary unit of South Africa.
[b]Silver and uranium are by-products of gold.
[c]Sales value of uranium production not disclosed.
[d]Coal means anthracite and bituminous coal. For mines producing coking coal, the number of persons in service and at work on surface may include persons working in coke plants.
[e]Production and sales value refer to metal and concentrates.
[f]Production and sales value refer to ore and sands.
[g]Production and sales value refer to ores only.
[h]Production and sales value refer to hematite and magnetite.
[i]Production and sales value not listed separately.
[j]Platinum and platinum-group metals (PGM) included under "other," but production and sales value of platinum and PGM not disclosed.
[k]Production and sales value not listed separately.

For occupational health and safety purposes, the South African minerals industry is treated as a single entity embracing underground mining, surface mining and metallurgical works. Use of surface mining for exploitation of coal, copper, iron ore, diamonds and other minerals has increased markedly in recent years, as has production of iron, steel and ferroalloys from metallurgical works. However, underground mining still represents by far the largest of the three branches, and since the risk of injury and the difficulties of hazard abatement are in general far greater in underground mines than on the surface, this chapter is confined to underground mining.

Respiratory diseases caused by inhalation of airborne particulates were first recognized in Witwatersrand gold miners at the turn of the century; subsequent inquiries clearly demonstrated the gravity of the hazard. Remedial measures applied in Witwatersrand mines eventually reduced the hazard considerably, and were later extended to all mines in South Africa. There undoubtedly still is a significant occupational health hazard in South African mines, but in most sectors of the industry risks associated with rockbursts, falls of hanging, heat stress, methane explosions, inrushes and similar, more immediate, hazards are usually accorded greater priority than the seemingly remote probability of diseases caused by the inhalation of respirable airborne particulates.

HISTORICAL REVIEW OF HEALTH HAZARDS

On the Central Witwatersrand, i.e., in the area where gold was first discovered, the auriferous quartz "banket" reefs outcrop on surface and dip steeply south, sandwiched between massive layers of quartzite. Early miners followed the narrow reefs (1-2 m thick) down-dip, and when the Second Anglo-Boer War broke out in 1899, the deepest workings had already reached a vertical depth of just over 1 km below surface. Initially, all blastholes were drilled manually, but in 1892 pneumatic drills were introduced for development work, and by 1899 more than 2000 drills were in use. With no water for dust suppression, no mechanical ventilation and multishift working, conditions underground are said to have been appallingly bad. Statutory health and safety regulations were enacted by the Zuid-Afrikaansche Republiek in 1893, but were apparently ignored by the industry, and the small inspectorate was powerless to enforce them.

Witwatersrand mines closed during the hostilities, but began to reopen in 1901. In his first postwar report, the Government Mining Engineer [1902] drew attention to the fact that 225 out of the 1377 white drill operators employed immediately before the war were known to have died from natural causes during the wartime closure. No positive suggestions were made to account for this grossly excessive mortality rate, but in the following year the Miners' Phthisis Commission [Weldon 1903] was appointed to inquire into "the disease commonly known as miners' phthisis."

The commission's investigations were fairly cursory—for example, no radiographs were taken—but limited medical examinations of 1210 of the 4403 white miners then working in Witwatersrand mines showed that a significant number had contracted silicosis; 92% of the men with silicosis had worked on pneumatic drills and, on average, these men had spent 6.5 years at the task. A few measurements of total dust concentrations in mine workings made for the commission are quoted in Table II.

No details are given of sampling methods (sugar-tubes were probably used) or on the selection of sampling sites, but to some extent the figures confirm subjective impressions of dust conditions. Boyd [1928], after quoting the same figures, mentions that "as much as 1500 mg/m^3" was measured in a drive in an East Rand mine.

In 1907 the Mining Regulations Commission [Krause 1910] was appointed to examine the existing regulations and suggest how they might be amended "for the better protection of the health and safety of persons working in mines." In its final report the commission emphasized the need for better ventilation, use of water for dust suppression, and measures to protect mineworkers from exposure to blasting fumes and dust. Shortly after political unification of the four provinces in 1910, the new Union parliament passed the Mines and Works Act 12/1911, which empowered the governor-general to promulgate regulations on a wide

Table II. Gravimetric Measurements of Total Airborne Dust in Witwatersrand Mines Made for Miners' Phthisis Commission, 1902 (Weldon 1903)

Workplace	Dust Concentration (mg/m^3)[a]
Face of Drive	423
Face of Drive	192
Face of Drive Behind Sprays	53
100 ft (30.5 m) from Face 10 min After Blast	82
Stope	14
Stope	32
Raise	165
End of Drive 5.5 hr after blast	42

[a]Commission's results quoted in gr/ft^3; 2289 gr/ft^3 = mg/m^3.

variety of matters affecting occupational health and safety in mines; shortly afterward, new regulations were published incorporating most of the commission's recommendations. Hand-in-hand with the new act to regulate working conditions in mines, Parliament enacted the Miners' Phthisis Allowances Act 34/1911—possibly the first statute anywhere in the world to provide compensation benefits for an occupational disease, and the forerunner of the far more complex Occupational Diseases in Mines and Works Act 78/1973 applicable today. Introduction of the new regulations, coupled with a policy of more stringent enforcement on the part of the mines inspectorate, gradually had the desired effect, as can be seen from the figures reproduced in Table III [Irvine et al. 1930].

The data in Table III were collected by a Chamber of Mines dust sampling unit set up in 1914. Initially, the unit only used the sugar-tube, and sampling and assessment techniques were not necessarily the same as might have been used in 1902 (see Table II); in particular, the unit used a technique introduced in 1915 to remove particles larger than 12 μ before weighing the sample [Boyd 1930]. The unit adopted the konimeter in 1919, but fortunately continued to use the sugar-tube for parallel sampling with the konimeter for several years. In 1916 the M&W Act regulations were amended to provide for the appointment of a dust inspector in all the larger Witwatersrand mines, and as a result the dust sampling unit's functions were gradually transferred to individual mines.

It is interesting to note that Irvine et al. [1930] considered the "danger mark" to be 5 mg/m^3, whereas 44 years later the National Institute for Occupational Safety and Health [NIOSH 1974] recommended that occupational exposure to airborne crystalline silica "shall be controlled so that no worker is exposed to a time-weighted average (TWA) concentration of free silica greater than . . . 0.050 mg/m^3 . . . as determined by a full-shift sample for up to a 10-hour workday, 40-hour workweek."

Table III. Gravimetric Measurements of Total Airborne Dust (Less Particles Larger than 12 μm) Made by the Chamber of Mines Dust Sampling Department, 1915–1927

	1915	1917	1919	1921	1923	1925	1927
General average (mg/m^3)	4.9	3.8	3.4	1.6	1.3	0.9	1.2
Development (mg/m^3)	6.9	5.4	3.5	2.3	1.9	1.0	1.2
Stopes (mg/m^3)	3.4	2.9	1.9	1.2	0.9	0.7	0.9
Ore Bins (mg/m^3)	4.4	4.2	2.9	2.1	1.8	1.6	1.9
Total Number of Samples	1758	6188	7491	6695	4872	2399	2869
Percent >5 mg/m^3	27.0	20.0	10.0	4.0	3.4	1.7	1.4

Early data on the chemical composition of in situ rock and airborne dust in Witwatersrand mines published by McEwan and Buist [1930] are reproduced in Table IV. Since 1930 the only published study on airborne dust in Witwatersrand mines seems to be a paper by Beadle and Bradley, [1970], in which the authors review quartz determinations by X-ray diffraction in samples from Witwatersrand mines. The authors comment:

> The mass concentration of total dust and of quartz as well as the percentage of quartz vary greatly from mine to mine and from one working place to another. This, to a lesser extent probably applies also to the same working place or mine sampled periodically.

The authors draw attention to the fact that the mined rock contains 60–80% of quartz, but that the quartz content of total dust was found to be only about 31%. They also found that the mass concentration of respirable dust generally represented 60–80% of the mass concentration of total dust.

By the 1930s, dust conditions in Witwatersrand mines had been transformed, and this favorable trend probably continued into the 1940s and 1950s. Changes in sampling and assessment techniques and a dearth of meaningful sampling data make it difficult to assess more recent trends, but there is some evidence [Martinson 1970] to suggest that

Table IV. Chemical Composition of in Situ Rock and Airborne Dust: Central Witwatersrand Mines, 1930[a]

	Country Rock, Hanging and Footwall (%)	Banket Reef (%)	Airborne Dust (%)
SiO_2	76.27	86.22	80.69
Al_2O_3	14.09	5.85	8.58
Fe_2O_3	4.71	1.97	0.50
FeS_2		2.19	[b]
CaO	0.23	0.99	2.78
MgO	1.64	0.78	1.40
Water	2.35	[c]	2.93[d]
Total	99.19[e]	97.91[e]	96.88[e]

[a]Means of "thousands" of sugar-tube samples (after HCl).
[b]Not given.
[c]Water not determined.
[d]Loss on ignition (includes organic matter).
[e]Alkalis not determined.

favorable trends were reversed in the 1950s. On the other hand, a Chamber of Mines [1980] report asserts that there has been "a significant decrease in average dust concentrations" since 1964.

Since the Miners' Phthisis Commission [Weldon 1903] and Mining Regulations Commission [Krause 1910] reported early in the century, various aspects of the health hazard in Witwatersrand mines have been studied by an extensive series of official commissions and committees. Many were preoccupied with compensation matters—long a continuous issue in South African politics—but more scientific studies inevitably have been hindered by the lack of relevant data on exposures and biological responses.

Thus far, reference has been made only to health hazards in Witwatersrand mines. It may be supposed that there has been a health hazard in all underground mines in South Africa since the start of mining operations—certainly all mines of any significance are now "controlled" mines in terms of the Occupational Diseases in Mines and Works Act—but with the possible exception of asbestos mines, the risk of health injury has never approached levels experienced in Witwatersrand gold mines. There are two explanations for this state of affairs: (1) standards of dust suppression and ventilation have historically been influenced greatly by Witwatersrand practice, and (2) other mineral dusts (again excluding asbestos) are in general not as pathogenic on a mass-for-mass basis as quartz dust. However, the adage that all dust is dangerous is still undoubtedly true.

As noted earlier, statutory compensation for mineworkers who contract occupational diseases was first introduced in South Africa in 1911. Initially, compensation was limited to white miners who contracted silicosis in Witwatersrand gold mines, but over the years the scheme has been extended to embrace virtually all mines and all mineworkers (and exmineworkers) who contract one or other of the specified compensable diseases; likewise, the scale of benefits has gradually increased in real terms since compensation was first introduced. It would be unrealistic to ignore the possibility that the scheme may become more expensive in future.

CURRENT STATUTORY PROVISIONS

Two statutes relate directly to health hazard in mines and works. The Mines and Works Act 27/1956 authorizes the state president to issue regulations dealing with working conditions in mines and works. The Occupational Diseases in Mines and Works Act 78/1973 provides for compulsory medical inspections for all persons who perform *risk work* (all work underground is classed as risk work), the payment of compensation to persons who contract compensable diseases as a result of performing risk work, and the collection of levies based on risk from controlled mines and works to pay for compensation benefits. Aspects of the Occupational Diseases in Mines and Works Act are considered elsewhere, and the rest of this section is devoted to Mines and Works Act regulations relating to airborne particulates and dust sampling.

Starting at the entrance to a mine ventilation system, regulation 10.6.1 provides that "[a]s far as practicable the ventilating air entering a mine shall be free from dust, smoke or other impurity," and the next paragraph stipulates that:

> 10.6.2 The workings of every part of a mine where persons are required to travel or work shall be properly ventilated to maintain safe and healthy environmental conditions for the workmen and the ventilating air shall be such that it will dilute and render harmless any inflammable or noxious gases and dust in the ambient air.

A later regulation directs that no person shall enter or remain in any part of a mine "if the air contains harmful smoke, gas, fumes or dust perceptible by sight, smell or other senses" unless the person is wearing an approved protective device.

Regulation 10.6.6 provides that under normal working conditions the concentration of certain specified gases in the general body of the air shall not exceed limits laid down in the regulation, and subparagraph f reads:

> (f) the concentration of dust shall not exceed such standard as may from time to time be specified by the Government Mining Engineer.

Apparently the Government Mining Engineer (GME) has on occasion issued standards through the mines inspectorate for different sectors of the industry—notably asbestos mining—but the standards are not published for general information and seemingly no mine or person has been prosecuted for contravention of the standards for many years.

Regulations specify minimum ventilation flowrates and velocities for different types of workings, and require appropriate ventilation measurements to be made and recorded at not more than monthly intervals in coal mines and three-month intervals in metal and diamond mines.

In terms of regulation 2.16.1, a certified ventilation officer must be appointed for every controlled metal and diamond mine where >1000 persons are employed underground in any one shift. In earlier regulations this official was referred to as a dust inspector, and even today his principal duty is to examine and report to the manager on:

(a) all matters relating to the mine's water supply, its quality, distribution and use,

(b) the condition of the necessary appliances for using water at each working place and elsewhere,

(c) the dust sampling of the mine, more particularly as regards development ends, and

(d) the condition of the mine relating to ventilation and health, more particularly as regards the amount of air supplied during the working shift, in all development ends and working places in which there is no through ventilation current.

Regulation 10.9.1 provides that where a ventilation officer is appointed in terms of regulation 2.16.1

determinations shall be made during the main working shift not less than once in three months of the ventilation and environmental conditions and the amount of dust in the air in the main airways and at the faces of working stopes, development ends and shafts in the course of being sunk, and such other places as directed by the manager.

In coal mines, dust sampling is related to the operation in progress:

10.9.4 In every coal mine measurements shall be made during the main working shift not less than once in six months or at such intervals as the Government Mining Engineer may permit of the amount of dust in the air in representative working places in each section while drilling, cutting, breaking, loading or transfer of coal or rock is taking place.

The Mines and Works Act regulations do not define "dust," and do not specify the type of instrument or strategy to be used for measuring "the amount of dust in the air," or by whom the sample should be taken. Likewise, no provision is made for approval of dust sampling instruments or checks on their efficacy. Several other regulations are relevant in relation to airborne particulates—notably those dealing with the supply and application of water for dust suppression and measures to prevent workers from being exposed to dust and fumes produced by blasting operations—but need not be examined in the context of this chapter.

SAMPLING INSTRUMENTS

A variety of sampling instruments have been used in South African mines at one time or another, but for routine sampling the konimeter has for many years been the standard sampling device in all mines other than coal mines, where the modified thermal precipitator (TP) is the standard instrument. The two instruments are discussed in some detail below, since they are not widely used elsewhere, and are followed by shorter comments on other instruments sometimes used in South Africa for special purposes.

Konimeter

The konimeter is a simple, hand-held sampling device in which the release of a spring-loaded plunger draws a 5-ml sample into the instrument through a narrow jet discharging at right-angles and in close proximity to a glass slide coated with a mixture of petroleum jelly, xylol and dioxane. Particles in the air sample are collected by impaction on the slide in the form of a "spot," and after each sample the glass slide is rotated a few degrees to bring a clear space on the slide opposite the jet; each slide can nominally accommodate 58 samples. The slide is removed from the instrument and treated (first ignition at 550°C, immersion in hot 50% hydrochloric acid, second ignition at 550°C) to eliminate carbonaceous matter and soluble salts before the spots are identified and counted under a microscope (150X) using dark-field illumination. A graticule with two 18° sectors is positioned over one spot at a time for counting purposes; only particles inside the two sectors are counted, and particles larger than 5 μm are excluded. Particles smaller than about 0.25 μm cannot normally be seen at the recommended magnification, and counts are converted to a number concentration per milliliter.

Standard techniques for sampling and assessing samples are detailed in "Routine Mine Ventilation Measurements" [Chamber of Mines 1972].

The konimeter was originally developed in 1916 by the (then) GME in his home workshop as a replacement for the cumbersome sugar-tube apparatus in use at the time in Witwatersrand mines [Kotze 1919]. The first version, known as the Kotze konimeter after its inventor, could only record one spot per slide, but Boyd [1928] describes an early multispot version capable of recording 58 spots (the circular konimeter) and Rees and Rabson [1948] introduced a number of further modifications (the Witwatersrand konimeter). More recently Quilliam and Kruss [1972] described the addition of a T-shaped size selector to the jet intake to serve as an elutriator, and a Chamber of Mines Report [1979] refers to the introduction of yet another improved instrument (the type-R konimeter). Individual authors furnish dimensions and other specifications, but neither the specifications nor the sampling characteristics of the konimeter have ever been defined or approved officially.

The South African konimeter and foreign versions of it have been used on an ad hoc basis in several countries, and even have been used for routine dust sampling in some overseas jurisdictions. Today, however, it is apparently only used for routine purposes in South Africa.

The advantages and disadvantages of the konimeter have been debated hotly on several occasions. Apart from the notorious difficulty of identifying individual spots correctly and the risk of mistakes, contamination and observer bias during treatment and counting, the major criticisms leveled at the konimeter include:

1. The fundamental objection, stressed by Beadle [1966] and others, is that the sampling period for one spot is less than a second, whereas it is well known that dust concentrations can change by orders of magnitude over relatively short periods. The recommended practice of recording three spots in succession at each sampling point and accepting the mean of the three as the sample value does not effectively counteract this objection;

2. Based on practical trials by Patterson [1937], Beadle [1951,1954a,1966] and others, the konimeter has a low collecting efficiency when sampling high concentrations of fine (<1-μ) dust, and the konimeter has an apparently high collecting efficiency when sampling low concentrations of coarse dust.

3. In 1952 Kerrich [1966] identified an inexplicable but statistically significant "wandering bias" between two apparently identical instruments when used for side-by-side sampling in a Witwatersrand mine. A standard overhaul of both instruments radically changed but did not eliminate the bias.

4. Balashov and Rendall [1959] suggested that deposition of particles in the jet when sampling high dust concentrations may contaminate subsequent samples.

Among ventilation practitioners, the konimeter has a number of dedicated supporters who argue that despite its known failings the instrument can, in the hands of a well-trained and conscientious observer, be used to good effect in dust control programs [van Nierop 1973]. In this context it may be noted that in the United States, criticism has also been leveled at newer, more sophisticated dust sampling instruments [Cook 1980; GAO 1975; NBS 1975].

Modified TP

As a result of mounting criticism of the konimeter [Beadle 1951], steps were taken in the late 1940s to find an acceptable replacement. At the time the original TP described by Green and Watson [1935] seemed to have several desirable features lacking in the konimeter (i.e., long sampling time, high collecting efficiency with fine dust, known sampling characteristics). Furthermore, in the immediate postwar period the standard TP was being used in several countries, including South Africa, to measure dust concentrations in mines for special purposes. On the other hand, the apparatus was bulky, not suitable for sampling on the move, required a fair degree of expertise to operate, and samples had to be counted laboriously under a microscope.

Kitto and Beadle [1952] described in some detail a modified TP (MTP) developed by the Chamber of Mines in response to this criticism, and Beadle [1954b] described a photoelectric assessor developed for use in conjunction with the MTP to replace optical counting. The sampling instrument was subsequently fitted with an impingement separator to simulate approximately the retention characteristics recommended at the 1959 Johannesburg Pneumoconiosis Conference [Jacobs 1968].

Field trials showed that the MTP was not suitable for sampling dust in gold mines, but with minor modifications [Craig 1960; Maxwell 1960] the two instruments were adapted for sampling coal dust and have been used successfully for this purpose since 1956. Kitson and Winer [1960] and Haven [1966] review the sampling system used by most coal mines in South Africa, and detailed operating instructions and sampling procedures are given in "Quality of Mine Air" [Chamber of Mines 1965].

The MTP samples continuously at 10 ml/min for a period selected by the observer ranging from 1 to 10 min. Air enters the instrument through a small horizontal elutriator and then passes through a narrow gap between a glass slide and a thin nichrome wire stretched transversely across the slide and 0.2 mm from it. The wire is backed by a refractory strip and is heated to 150°C with power from a caplamp battery. Power for the aspirator is drawn from the same source. Respirable particles entering the zone of influence of the hot wire migrate toward and adhere to the comparatively cool surface of the slide. At the end of the selected sampling period, the device is automatically switched off and the slide is advanced by a rack-and-pinion arrangement to a "clean" section of the slide. Normally, a 10-min sampling period is used, but longer or shorter periods can be used when lighter or heavier dust concentrations are anticipated. A maximum of 12 samples and a clearing strip can be accommodated on one slide. The instrument has been approved as being intrinsically safe by the GME. Slides are dried in a desiccator before being placed in the photoelectric assessor, and are also subjected to an ignition/immersion/ignition treatment if the ash content is required [Kitson et al. 1969].

The photoelectric assessor consists of a lightproof box in which two matched beams of light are beamed diagonally at photoelectric cells mounted at the far end of the box, so that the beams cross one another just in front of a traversing cradle supporting a single slide standing vertically on a long edge. The cradle incorporates a captive nut on a threaded spindle, which enables a slide to slowly traverse across the beams by rotation of the spindle. When the beams pass through a clean MTP slide, the output of the photoelectric cells can be balanced by a potentiometer to give a null deflection on a galvanometer, but when a slide with sample strips is inserted, the slide can be so arranged that one beam passes through a "clean" section of the slide while the other passes through a strip of dust deposited by the hot wire during sampling. The diminished intensity of the one beam reduces the output of the photocell in question and the magnitude of this imbalance is measured on a galvanometer. Two readings (one for each beam) are obtained from each sample strip and, after the mean of the two readings has been adjusted for sample size and assessor characteristics, the resulting numbers are quoted as dimensionless photoelectric readings (PER). A slide bearing 12 samples can normally be assessed in 15 minutes.

Based on side-by-side sampling of Witwatersrand gold-mine dust using standard TP operated by trained personnel, Beadle and Kerrich [1955] reported that the overall coefficient of variation of a single TP sample result is about 13%, but an analysis by Joffee and Williams [1963] of PER values derived from a similar series of tests using the MTP to sample coal dust indicated that the coefficient of variation of an individual PER value is likely to exceed 13%.

PER values are said to be proportional to the surface area of the dust sampled [Beadle 1954b; Kitto and Beadle 1952] and have apparently been used to good effect for dust control purposes in collieries [Kitson and Haven 1968].

MRE/113A (Isleworth) Gravimetric Sampler

The MRE/113A was developed in Britain at the Mining Research Establishment of the NCB in the early 1960s by Dunmore et al. [1964], and has for several years been used as the standard dust sampling instrument in British coal mines. It also has the unusual distinction of being specified in the federal Coal Mine Health and Safety Act of 1969 as the reference instrument for defining the maximum allowable concentration of coal dust in U.S. coal mines. The instrument has been used sporadically in South Africa, and Kitson [1970], for example, quotes results from parallel sampling with the MTP and MRE/113A.

The MRE/113A has been described on many occasions, but in essence it is a full-shift, area sampler and is designed to sample continuously at 2.5 liter/min for 8 hr. Air enters the instrument through a horizontal, parallel-plate elutriator, and respirable particles passing through the elutriator are deposited on a membrane filter, which is removed for weighing and analysis of the dust. The retention characteristics of the elutriator are based on aerosol penetration characteristics defined by the British Medical Research Council. The instrument is equipped with an internal power supply, and is intrinsically safe. The performance of the MRE/113A has been tested against other sampling instruments in several parts of the world, notably against personal samplers in the United States [Harris et al. 1976; Tomb et al. 1973].

Personal Samplers

The title "personal sampler" is nonspecific, but is generally taken to mean one of the several sets of equipment approved in terms of the federal Mine Safety and Health Act of 1977, or comparable equipment. As the name implies, the complete sampler is carried on the person of a mineworker, generally to measure full-shift personal exposure.

The sampling head, usually pinned to the subject's clothing as near head-height as possible, incorporates a cassette containing a membrane filter, the outlet side of which is connected by a flexible tube to a pump-and-battery unit worn on the subject's belt, while the intake side is connected to a 10-mm nylon cyclone elutriator. The pump draws air through the sampling head at a flowrate of 2 liter/min; at this flowrate the elutriator has a retention characteristic defined by the penetration curve for respirable aerosols published by the American Conference of Government Industrial Hygienists.

Tests show that approved personal samplers have an intrinsic variability of about 7%, but when used in mine by trained scientists the uncertainty could be as high as 50% [NBS 1975]. Any personal sampling device is, of course, also susceptible to rough treatment, which may have a marked effect on sampling results.

Instantaneous Dust Measuring Instruments

A chronic disadvantage of the sampling instruments considered up to now is the inevitable delay between taking the sample and obtaining the result. At best this may be a matter of hours or at worst a matter of days, but in either event the atmospheric environment may change radically in the interim; for control purposes the result is largely irrelevant by the time it becomes available. In the last decade a number of instruments have been developed for making rapid measurements of respirable dust in mine air using physical phenomena such as β-radiation attenuation, electrostatic precipitation coupled to a piezoelectric balance and scattering of a beam of light. Despite the obvious advantages of such instruments for dust control programs and exposure measurements, high initial cost and maintenance difficulties associated with remoteness have thus far inhibited their widespread use in South African mines. However, of the few instruments in use, Simslin II may warrant a few brief remarks.

Simslin is named after the British research establishment in which it was developed (safety in mines scattered light instrument). Infrared radiation emitted by a small laser diode is focused onto a column of air drawn into the instrument through a horizontal elutriator at a flowrate of 625 ml/min, and a fraction of the incident light scattered between 12 and 20° to the forward direction by particulates in the sampled air is focused onto a detector to produce a signal proportional to the dust concentration in the column. The detector output is fed through appropriate circuitry to a digital LCD that is updated every second; a second display shows the cumulative average respirable dust concentration in mg/m³, updated every 15 minutes. Data from the instrument can be fed into three separate channels simultaneously, namely:

1. conventional chart recorder;
2. data transmission line;
3. internal digital recording system (SIMSTOR).

After passing through the cylindrical photometer unit, respirable dust in the sample stream is collected on a filter for weighing and analysis. Since the filter collects the same dust that causes the scattered light reading, comparison of the two values provides a simple method of calibrating the photometer. Furthermore, as the retention characteristics of the SIMSLIN and MRE/113A elutriators are matched, the two instruments are ostensibly compatible. Simslin and other advanced dust sampling instruments have most of the technological features needed to revolutionize dust sampling in mines, but few, if any, dust sampling systems presently are sufficiently sophisticated to fully use the technological features.

SAMPLING STRATEGY

Beadle [1960] remarked "[i]t is obvious that before one starts dust sampling one should be quite clear as to the purpose of the sample," and stresses that the purpose of sampling greatly affects the way samples are taken and assessed, and the desired accuracy. Beadle lists eight "possible objects" of sampling, but seemingly these could be reduced to two:

1. to provide quantitative feedback on the quality of the mine atmospheric environment for positive engineering control of atmospheric quality by airflow regulation, air conditioning, the use of barriers or a combination of these measures; and/or
2. to enable assessment of the exposure of mining personnel to airborne particulates.

A sampling system satisfying the first requirement probably also would provide the data needed for assessing exposure.

Despite the obvious need to establish a valid rationale for sampling aerosols in mines, this aspect of sampling has, paradoxically, received little attention in the ventilation literature. The respective merits of area and personal sampling have been debated at some length, but discussion has centered on instrumentation and compliance criteria rather than strategic considerations. In South African mines, at least five different sampling systems can be identified.

Routine Sampling in Gold and Most Noncoal Mines

Sampling procedures used in Witwatersrand gold mines today were established in the 1920s and 1930s following the introduction of the konimeter, and since World War II, Witwatersrand practice has been adopted in most South African mines, except coal mines. Typically an observer—usually a full-time ventilation official employed by the mine—identifies a number of sampling stations in relation to each working place (notably intake, face and return) and at each station takes a set of three konimeter spots at one-minute intervals. The observer makes a note of any activity in the vicinity of the sampling station, but takes the sample with conditions "as found."

Back on the surface, the konimeter slide is treated and counted by the observer or, in some larger mines, a microscopist. Means of the three spots at each sampling station and other environmental information are used by the observer to compile a ventilation report on each workplace. The report is scrutinized by the ventilation officer and line-management officials with statutory responsibility for the working place, and is then filed.

Routine Dust Sampling in Coal Mines

A large majority of South African coal mines are members of the Chamber of Mines and make use of dust sampling services provided by the Chamber since 1956. Other coal mines make their own arrangements.

For routine surveys, the Chamber of Mines coal mine sampling units use the MTP and photoelectric assessor described above. During periodic visits to a mine, an observer from the unit visits each working section in turn to take dust samples and record other environmental measurements. Normally, samples are taken in intake and return airways, and in the vicinity of workmen performing their usual tasks during periods of maximum dust production associated with standard mining operations such as cutting, drilling and loading. The normal sampling time for the MTP is 10 min, but if an operation is stopped temporarily for any reason during the sampling period, sampling is interrupted. MTP slides are treated and assessed at regional dust sampling laboratories, and the results are incorporated in a full ventilation report on each section submitted to the mine concerned [Haven 1966; Kitson and Winer 1960]. Photoelectric assessors are calibrated against standard slides at frequent intervals, and a small percentage of samples are assessed under a microscope as a check on PER values.

Sampling to Measure Full-Shift Exposure

Surveys occasionally are undertaken to estimate dust exposures for epidemiological studies of work-related respiratory diseases, and in this connection a survey conducted by Beadle and his colleagues in Witwatersrand gold mines in the late 1950s remains a noteworthy example (reported by Page-Shipp and Harris [1972]). During the survey, a total of 650 white mineworkers were followed portal-to-portal by highly trained observers carrying and operating three sampling instruments:

1. standard TP to provide a continuous, full-shift sample assessed by a microscopist in terms of respirable surface area;
2. MTP, providing a series of discrete 10-min samples throughout the shift and evaluated in the PER; and
3. konimeter used to take a single spot at 10-min intervals throughout the shift and assessed in the usual fashion to give number concentrations.

Subjects were chosen randomly from 11 different occupations with due regard for the total number of men in each occupation in all Witwatersrand mines and exposure patterns developed during the survey. Beadle et al. [1970] quoted dose-response relationships derived from the survey and medical histories of a cohort of mineworkers.

Sampling to Determine Risk

Section 18 of the Occupational Diseases in Mines and Works Act 78/1973 provides for the establishment of a statutory risk committee whose primary function is to determine the risk (expressed as a percentage) of contracting a compensable disease as a result of performing risk work in controlled mines and works. The percentage risk is mainly based on measured dust levels and the pathogenicity of the various mineral dusts, and is used to assess the levy each controlled mine and works must pay toward the total cost of benefits paid to workers who contract compensable diseases. The committee uses various criteria for determining dust levels in different sectors of the minerals industry, but for coal mines participating in the Chamber of Mines dust sampling scheme the committee uses routine MTP sampling results (PER) for risk determination.

For Witwatersrand gold mines, the committee bases its risk determinations on dust levels measured by another Chamber of Mines dust sampling unit established in response to the 1973 Act. In this case an observer from the unit visits randomly selected sections of each mine during the main working shift and takes single konimeter spots in the vicinity of every third workman he encounters during a detailed tour of the section. The konimeter slides are treated and counted at a central laboratory in accordance with standard procedures, and the results forwarded to both the mine and the risk committee. A Chamber of Mines Report [1979] mentions the possibility of replacing the konimeter with personal gravimetric dust samplers for risk sampling.

Sampling by the Mines Inspectorate

In terms of section 2 of the Mines and Works Act 27/1956 the GME is responsible for the supervision of all mines, works and machinery; section 3 provides that this responsibility is exercised in mines by a mines inspectorate. In the course of periodic visits to mines, inspectors use konimeters to take dust samples at various defined points in the ventilation system; the samples are assessed using standard procedures, and annual averages of dust concentrations at the various sampling points are published in the GME annual report.

The GME office also maintains an air quality section equipped with a range of sampling devices for undertaking risk sampling in special circumstances and other investigations.

CONCLUSIONS

Routine dust sampling has played a major role in abating health hazards in South African mines; in the light of probable technical and social changes in the mining industry, the importance of this role is likely to increase rather than diminish in the future. Sampling systems in use have in general developed in a pragmatic manner and, like sampling systems in other jurisdictions, could no doubt be improved and made more cost-effective. In this respect four matters seem to warrant attention:

1. The rationale underlying any system for sampling aerosols in mines should be established in explicit terms, since this dominates every aspect of system design. There is, of course, no unique formulation of sampling rationale, but usually it will relate specifically to engineering control of pathogenic aerosols in the mine atmosphere.
2. The overall uncertainty of a single sampling result is a function of (1) the intrinsic accuracy of the sampling instrument in question; (2) the manner in which the instrument is handled, particularly while sampling; (3) the intrinsic accuracy of the assessment procedure; (4) the degree

of care exercised by the person assessing the sample; and (5) the temporal and spatial variability of the sampled aerosol. All of these areas require more thorough study than they have received in the past, but it seems safe to say that there are sampling instruments commercially available today whose intrinsic accuracy and utility is superior to the instruments presently used for routine sampling.

3. For sampling and control purposes there is an urgent need for a better understanding of the factors affecting the formation, dispersal and decay of aerosols in mine workings; this, in turn, calls for a greater flow of relevant sampling data and more sophisticated analysis of sampling results. In some jurisdictions there is a marked reluctance to release any sampling data, and clearly this attitude has helped to stifle meaningful analysis.

4. Any rational dust control program—and, indirectly, any dust sampling program—must ultimately relate to the risk of injury and the social acceptability of risk. Dose-response relationships have, of course, been published for various mineral dusts in a number of different countries, but these are mostly based on radiological evidence of injury and a single characteristic of the aerosol. However, an increasing number of mineworkers are receiving compensation in circumstances where radiological evidence is inconclusive or absent, and it seems reasonable to suggest that a combination of aerosol characteristics would predict injury more accurately. Further epidemiological studies are needed to clarify these matters.

REFERENCES

Balashov, V., and R. Rendall (1959) "Some Remarks on the Performance of the Witwatersrand Konimeter," *J. Mine Vent. Soc. SA* 12:1-4.

Beadle, D. G. (1951) "An Investigation of the Performance and Limitations of the Konimeter," *J. Chem. Met. Min. Soc. SA* 51:265-283.

Beadle, D. G. (1954a) "The Performance and Improvement of Dust Sampling Instruments," MSc Thesis, Rhodes University, Grahamstown.

Beadle, D. G. (1954b) "A Photo-electric Apparatus for Assessing Dust Samples," *J. Chem. Met. Min. Soc. SA* 55:30-39.

Beadle, D. G. (1960) "Objects of Dust Sampling," in *Proceedings of the Pneumoconiosis Conference, Johannesburg, 1959*, A. J. Orenstein, Ed. (London: Churchill), p. 4.

Beadle, D. G. (1966) "Contribution to J. E. Kerrich: Note on the Performance of the Witwatersrand Konimeter," *J. SA Inst. Min. Met.* 66:402-407.

Beadle, D. G., and A. A. Bradley (1970) "The Composition of Airborne Dust in South African Gold Mines," in *Pneumoconiosis—Proceedings of the International Conference, Johannesburg, 1969*, H. A. Shapiro, Ed. (Oxford: Oxford University Press).

Beadle, D. G., and J. E. Kerrich (1955) "A Statistical Examination of the Performance of the Thermal Precipitator," *J. Chem. Met. Min. Soc. SA* 56:219-239.

Beadle, D. G., E. Harris and G. K. Sluis-Cremer (1970) "The Relationship Between the Amount of Dust Breathed and the Incidence of Silicosis:

An Epidemiological Study of South African European Gold Miners," in *Pneumoconiosis—Proceedings of the International Conference, Johannesburg, 1969*, H. A. Shapiro, Ed. (Oxford: Oxford University Press), pp. 473-477.

Boyd, J. (1928) "The Estimation of Dust in Mine Air on the Witwatersrand," *J. SA Inst. Eng.* 26:142-160.

Boyd, J. (1930) "The Estimation of Dust in Mine Air," in *Proceedings—Third Empire Mining and Metallurgical Congress, South Africa, 1930*, J. A. Vaughan, Ed. (Johannesburg, SA: Congress Offices).

Chamber of Mines (1965) "Quality of Mine Air," Johannesburgh, SA.

Chamber of Mines (1972) "Routine Mine Ventilation Measurements," Johannesburg, SA.

Chamber of Mines (1979) "Chamber of Mines of South Africa—Research and Development Annual Report 1979," Johannesburg, SA.

Chamber of Mines (1980) "Chamber of Mines of South Africa—Research and Development Annual Report 1980," Johannesburg, SA.

Cook, N. G. W., Chairman (1980) "Report of the Committee on the Measurement and Control of Dust," NMAB-363, National Materials Advisory Board, National Academy of Science, Washington, DC.

Craig, D. K. (1960) "Photoelectric Assessors," *J. Mine Vent. Soc. SA* 13:45-56.

Dunmore, J. H., R. J. Hamilton and D. Smith (1964) "An Instrument for the Sampling of Respirable Dust for Subsequent Gravimetric Assessment," *J. Sci. Instr.* 41:669-672.

GAO (1975) "Report to the Congress by the Comptroller General of the United States: Improvements Still Needed in Coal Mine Dust-Sampling Program and Penalty Assessments and Collections," RED-76-56, General Accounting Office, Washington, DC.

Government Mining Engineer (1902) "Transvaal Mines Department—Half-Yearly Report of the Government Mining Engineer for the 6-Months Ending December 31st 1901," Government Printing Works, Pretoria, SA.

Green, H. L., and H. H. Watson (1935) "Physical Methods for the Estimation of the Dust Hazard in Industry," Special Report No. 199, Medical Research Council, London.

Harris, H. E., W. C. Desieghardt and L. A. Riva (1976) "Development of an Area Sampling Methodology," OFR 105-79, NTIS PB 299926, Bureau of Mines, Department of the Interior, Washington, DC.

Haven, Y. J. F. (1966) "The Technique of Routine Airborne Dust Sampling on Coal Mines in the Transvaal and Orange Free State," *J. Mine Vent. Soc. SA* 19:81-85.

Irvine, L. G., A. Mavrogordato and H. Pirow (1930) "A Review of the History of Silicosis on the Witwatersrand Goldfields," in *Silicosis—Records of the International Conference, Johannesburg 1930*, Studies and Reports F12 (Geneva, Switzerland: International Labour Office), pp. 178-208.

Jacobs, D. J. (1968) "An Impingement Separator for Sampling Respirable Gold Mine Dust with the Modified Thermal Precipitator," *J. Mine Vent. Soc. SA* 21:173-179.

Joffee, A. D., and C. D. Williams (1963) "A Statistical Examination of the Photoelectric Assessment of Dust Samples Taken with the Modified Thermal Precipitator," *J. Mine Vent. Soc. SA* 16:173-188.

Kerrich, J. E. (1966) "Note on the Performance of the Witwatersrand Konimeter," *J. SA Inst. Min. Met.* 66:219-226.

Kitson, G. H. J. (1970) "Calculation of the Relationship Between Photo-electric Readings and Particle Number, Area and Weight Concentrations in South African Coal Mine Dust Clouds," in *Pneumoconiosis—Proceedings of the International Conference, Johannesburg, 1969*, H. A. Shapiro, Ed. (Oxford: Oxford University Press), pp. 587-592.

Kitson, G. H. J., and Y. J. F. Haven (1968) "Photoelectric Readings: Their Significance for Dust Control on Collieries," *J. SA Inst. Min. Met.* 69:130-138.

Kitson, G. H. J., and P. Winer (1960) "Routine Airborne Dust Sampling in Collieries," *J. Mine Vent. Soc. SA* 13:153-158.

Kitson, G. H. J., Y. F. J. Haven and A. A. Bradley (1969) "Ash and Silica Content of Airborne Dust in Transvaal and Orange Free State Collieries," *J. Mine Vent. Soc. SA* 22:97-105.

Kitto, P. H., and D. G. Beadle (1952) "A Modified Form of Thermal Precipitator," *J. Chem. Met. Min. Soc. SA* 52:284-306.

Kotze, R. N., Chairman (1919) "Final Report of the Miners' Phthisis Committee," Government Printing and Stationery Office, Pretoria, SA.

Krause, F. E. T., Chairman (1910) "Transvaal—Final Report of the Mining Regulations Commission," Government Printing and Stationery Office, Pretoria, SA.

Martinson, M. J. (1970) In: *Pneumoconiosis—Proceedings of the International Conference, Johannesburg, 1969*, H. A. Shapiro, Ed. (Oxford: Oxford University Press), pp. 480-482.

Maxwell, D. K. (1960) "Production of Permanent Calibration Slides," *J. Mine Vent. Soc. SA* 13:205-207.

McEwen, A. F., and J. Buist (1930) "The Nature and Source of Dust in Mine Air," in *Silicosis—Records of the International Conference, Johannesburg, 1930*, Studies and Reports F13 (Geneva, Switzerland: International Labour Office), pp. 129-140.

Mining Statistics 1979 (1980) RP 96/1980 (Pretoria, SA: Government Printer).

NBS (1975) "Final Report to the Senate Committee on Labor and Public Welfare: An Evaluation of the Accuracy of the Coal Mine Dust Sampling Program Administered by the Department of the Interior," National Bureau of Standards, Washington, DC.

NIOSH (1974) "Criteria for a Recommended Standard: Occupational Exposure to Crystalline Silica," Publication No. NIOSH-75-120, National Institute for Occupational Safety and Health, PHS, Department of Health, Education and Welfare, Washington, DC.

Page-Shipp, R. J., and E. Harris (1972) "A Study of the Dust Exposure of South African White Gold Miners," *J. SA Inst. Min. Met.* 73:10-24.

Patterson, H. S. (1937) *Some Studies of Dust in Relation to Silicosis* (Johannesburg, SA: Hortors).

Quilliam, J. H., and J. A. L. Kruss (1972) "The Performance of Standard Konimeters Fitted with Size Selectors," *J. Mine Vent. Soc. SA* 25:60-68.

Rees, J. P., and S. R. Rabson (1948) "An Improved Type of Konimeter," *J. Chem. Met. Min. Soc. SA* 49:85-88.

Tomb, T. F., H. N. Treaftis, R. L. Mundell and P. S. Parobeck (1973) "Comparison of Respirable Dust Concentrations Measured with MRE and Modified Personal Gravimetric Sampling Equipment," RI 7772, Bureau of Mines, Department of the Interior, Washington, DC.

van Nierop, J. A. (1973) "Routine Dust Sampling," *J. Mine Vent. Soc. SA* 26:101-112.

Weldon, H., Chairman (1903) "Transvaal Colony—Report of the Miners' Phthisis Commission 1902-03," Government Printing and Stationery Office, Pretoria, SA.

CHAPTER 4

WORK-ENVIRONMENT MEASUREMENT LAW AND ITS PRESENT STATUS OF APPLICATION IN JAPAN

Katsunori Homma

National Institute of Industrial Health
Ministry of Labor
Kawasaki, Japan

ABSTRACT

On May 1, 1975, the Japanese government promulgated the Work-Environment Measurement Law. Briefly, this law is an imperative to maintain the quality of the working environment at a specified level and to endeavor to create a better and more comfortable working environment. In this law, the following subjects and items are defined and explained:

1. design, sampling and analysis relating to work-environment measurement;
2. workplaces subject to measurement of work environment;
3. measuring equipment, recording and preservation of measured data;
4. system of qualifications for work-environment measurement experts;
5. implementation of work-environment measurement; and
6. standards for work-environment measurement.

Since this law was promulgated, about 14,000 work-environment measurement experts have been certified by passing the national examination. They have acted according to prescribed methods in the law and have created many comfortable working environments.

BACKGROUND OF THE WORK-ENVIRONMENT MEASUREMENT LAW

The fundamental law for occupational health in Japan is the Industrial Safety and Health Law, which was enacted in 1972. Article 65 of this

59

law provides that the employer shall take the necessary measurements in the environment and keep the record of the results in those workshops prescribed by Cabinet order. Workplaces designated for measurement of the work environment by article 65 are shown in Table I.

On May 1, 1975, the Japanese government promulgated the Work-Environment Measurement Law. Briefly, the purposes of this law are to secure the optimum work environment by establishing the qualifications for work-environment measurement experts and regulating the affairs of work-environment measurement agencies, thereby to maintain the health of workers.

DEFINITIONS IN THE WORK-ENVIRONMENT MEASUREMENT LAW

Designated Unit Workplace

The concept of the unit workplace is introduced to clarify what should be controlled in the work environment. The unit workplace is defined as the area in which the exposure concentration is presumed not to fluctuate greatly nor to be random with regard to each worker if the average worker's exposure concentration fluctuates. Therefore, the designated unit workplace is not a compartment divided by building structure or work organization, but the compartment in which the concentration of harmful substances determined by the limited sampling method reflects the mean exposure concentration of workers. An example of designated unit workplaces is shown in Figure 1.

Table I. Workplaces Designated for Measurement of the Work
Environment by Article 65

Indoor working places generating extremely heavy noise
Indoor working places subjected to extremely high or low temperature or high relative humidity
Working places vulnerable to oxygen deficiency
Indoor working places where workers are engaged in specific organic solvent operations
Indoor working places with heavy generation of dust from earth, stones rocks, minerals or carbon[a]
Working places in underground galleries or tunnels[a]
Indoor working places where special chemical substances are manufactured or handled[a]
Indoor working places where workers are engaged in the specific operations with lead[a]
Working rooms where radioactive substances are handled[a]

[a]May contain aerosols.

A-Measurement and B-Measurement

Two basic ways of evaluating the work environment can be distinguished. A-measurement is evaluation through measuring the general air concentrations by area monitoring. B-measurement is evaluation of the worker's exposure concentration by a personal sampling device.

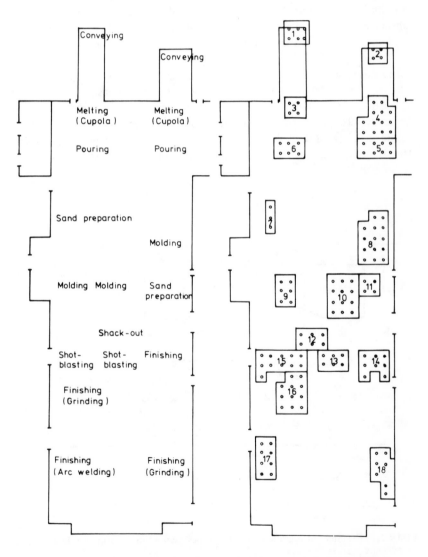

Figure 1.   Working process (left) and unit workplaces (right) in an iron foundry.

Sampling Point

The sampling points for A-measurement in the designated unit workplace should be five or more randomly determined points. However, if the unit workplace is remarkably small and there is a relatively uniform concentration of harmful substance, measurements should be made five or more times at a point in the worker's breathing zone. For the B-measurement point, one point should be so determined that, judging from the manufacturing processes, work habits and diffusion of harmful substances in the unit workplace as observed during the A-measurement, workers seem to be exposed to the maximum concentration in the unit workplace.

An example of determining sampling points is shown in Figure 2. That is, the center of the unit workplace is regarded as the origin, and from this point, lines are drawn crosswise at nearly equal intervals. Generally, for the A-measurement points, between 20 and 30 points should be selected, and this is a reasonable number to sample in a half-day. The B-measurement point is selected from places near generating points of harmful substances.

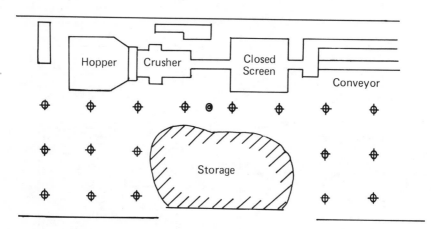

⊕ : Sampling point for A-measurement

◉ : Sampling point for B-measurement

Figure 2. Example of sampling points in designated unit workplace in the rock-crushing process.

Sampling Time and Period

The measurement should be conducted while work is in constant operation, and it should be planned so that sampling in the unit workplace is finished during a day's working time. Also, measurements should be carried out on two days. The sampling time should be more than 10 min at each sampling point of A-measurement. The B-measurement should be carried out within the A-measurement time, and its sampling time should be >15 min. The sampling period should correspond to one of the normal work. Generally, the period between the beginning and end of the A-measurement should be more than 1 hr.

Work-Environment Measurement Expert

Work-environment measurement experts are divided into two classes. Class 1 experts are permitted to conduct design, sampling and analysis only at specified workplaces, such as a mineral dust-generating workplace (Class 1: Mineral Dust), a beryllium- and/or cadmium-handling workplace (Class 1: Metal), or an α-naphthylamine-manufacturing workplace (Class 1: Special Chemical Substance). Class 2 experts are authorized to conduct design, sampling and analysis (limited to measurement using simple measuring instruments such as gas detecting tubes and light-scattering aerosol monitors) at all workplaces. Since this law had been promulgated, about 14,000 work-environment measurement experts have been certified by passing the national examination. The number of experts in each category is shown in Table II.

Table II. Number of Work-Environment Measurement Experts

Year	Class 1 Experts					Class 2 Experts
	Mineral Dusts	Metals	Special Substances	Organic Solvent	Radio-isotope	
1975	571	339	684	673	36	629
1976	505	509	761	1508	147	1140
1977	1142	425	794	1316	67	639
1978	218	255	588	683	65	569
1979	490	111	388	484	28	402
1980	387	137	266	441	29	322
1981	261	22	247	250	29	176
Total	3574	1798	3728	5355	401	3877

Work-Environment Measurement Agency

The work-environment measurement agency is defined as a person registered with the Minister for Labor or Chief of Prefectural Labor Standards Office and engaged in the measurement of work-environment of workplaces on commission basis at the request of others. To be registered with the Minister for Labor or Chief of Prefectural Labor Standards Office, the application submitted by the agency is required to fulfill the specified standards and contain none of the specified disqualifying factors. Further, the kinds of workplaces that the applying agency covers are specified in the registration.

Standards for Work-Environment Measurement

Measurement of the work environment shall be performed in conformance with the Standards for Work Environment Measurement established by the Minister for Labor, and the Minister for Labor is required to issue Guiding Principles in Work Environment Measurement for the purpose of adequate and efficient implementation of the measurement work. The following are the main items covered by the standards:

1. method of establishing measuring point,
2. sampling time of air specimen,
3. method of collecting air specimen,
4. method of analyzing the specimens, and
5. measuring instrument.

MEASUREMENT FOR CONCENTRATION OF MINERAL DUST

The concentration of airborne dust from earth, stones, rocks, minerals or carbon (excluding asbestos) shall be measured according to one of the following methods:

- collection method by filtration by using size selector and gravimetric analysis, or
- relative concentration indicating method (this is limited to the case where the above method is used simultaneously at more than one point in the unit workplace).

At present, light-scattering aerosol monitors, piezoelectric microbalance aerosol monitors, etc., are generally used. The size selector shall be one whose performance is illustrated by Figure 3 or others that give the same value.

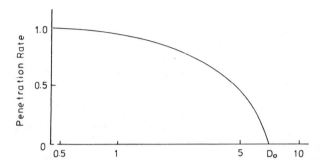

$$P = 1 - \frac{D^2}{D_o^2} \quad (D \leq D_o)$$

$$P = 0 \quad (D > D_o)$$

Figure 3.   Characteristic performance of size selector specified in the standards. P = penetration rate; D = aerodynamic particle diameter; D_0 = 7.07 μm.

CALCULATIONS FOR MEASURED VALUES
IN THE UNIT WORKPLACE

Concentrations of harmful substances in the unit workplace are expressed as a geometric mean concentrations (M_g) and geometric standard deviation (σ_g), because the distribution of the concentration of harmful substances in the air of a workplace is not normal, but log-normal. Thus, measured values on the first day at A-measurement:

$$C_{11}, C_{12}, C_{13}, \ldots, C_{1n_1}$$

the same on the second day:

$$C_{21}, C_{22}, C_{23}, \ldots, C_{2n_2}$$

Calculate two statistical values, after logarithmic change of C_{ij}.

$$\text{geometric mean: } M_g = 10^{1/2(\overline{X}_1 + \overline{X}_2)} \tag{1}$$

where

$$X_{ij} = \log C_{ij}$$

$$\overline{X}_1 = \frac{1}{n_1} \sum_{j=1}^{n_1} X_{1j}$$

$$\overline{X}_2 = \frac{1}{n_2} \sum_{j=1}^{n_2} X_{2j}$$

Geometric standard deviation, including the variances in the atmosphere and the day:

$$\log \sigma_g = \sqrt{\frac{\log^2 \sigma_1 - \log^2 \sigma_2}{2} + \log^2 \sigma_D} \tag{2}$$

where

$$\log^2 \sigma_1 = \frac{1}{n_1 - 1} \sum_{j=1}^{n_1} (X_{1j} - \overline{X}_1)^2$$

$$\log^2 \sigma_2 = \frac{1}{n_2 - 1} \sum_{j=1}^{n_2} (X_{2j} - \overline{X}_2)^2$$

$$\log^2 \sigma_D = \frac{1}{2} (\overline{X}_1 - \overline{X}_2)^2$$

EVALUATION FOR WORKING ENVIRONMENT

The effects of harmful airborne substances on a worker's health depend on the concentration levels, and so environmental control of individual concentration levels shall be required. To control the working environment, it is desirable to divide the working environment into at least three control sections. The first control section is the working environment that is expected to be sufficiently controlled. The third control section is one which needs to be improved. The second is an intermediate situation. The first control level is defined as the concentration level that distinguishes the first control section from the second; the second control level distinguishes the second control section from the third.

ESTABLISHMENT OF CONTROL LEVEL

The first control level is determined from:

$$\log E = \log Mg + 1.645 \log \sigma g \qquad (3)$$

It is indicated as "Line A" in Figure 4. The horizontal axis is the ratio of geometric mean concentration to the controlling concentration (Mg/E), and the vertical axis is the geometric standard deviation (σg).

The second control level is calculated from:

$$\log E = \log Mg + 1.151 \log^2 \sigma g \qquad (4)$$

"Line B" in Figure 4 shows the second control level.

The first control level for B-measurement is equal to the controlling concentration E, and the second control level for B-measurement is 1.5E.

PRESENT STATUS OF APPLICATION FOR WORK-ENVIRONMENT MEASUREMENT LAW

About 14,000 work-environment measurement experts, who are employed in their manufacturing companies or in about 300 work-environment

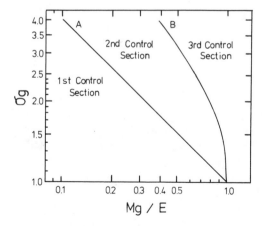

Figure 4.   Control levels and sections. A = first control level; B = second control level.

measurement agencies, have carried out sampling and/or measurement of harmful substances in work environments. Measured data should be recorded on standard forms and should be kept indefinitely.

Using the results of work environment measurement, it is possible to decide what control section corresponds to a measured workplace. For example, if first control section status is measured twice in succession at six-month intervals, then some part of the administrative regulation might be lightened. However, if third control section status is determined by work-environment measurement, reforming action should be carried out throughout the duration of that status.

Recently, the Japanese Occupational Safety and Health Administration has determined the status of working environment measurement in a selection of 130 workshops. Of those, 122 workshops have carried out work-environment measurements mostly by the work-environment measurement agencies. Furthermore, it appeared that 57.3% of workshops with mineral dusts belonged to the third control section, and only 20.1% belonged to the first control section.

CHAPTER 5

MEASUREMENT CONTROL OF RESPIRABLE DUST IN THE MINE ENVIRONMENT IN JAPAN

T. Tajiri, M. Kohno, A. Omori and S. Kato
National Research Institute for Pollution and Resources
Agency of Industrial Science and Technology
Ibaraki, Japan

ABSTRACT

This chapter reports on the status of respirable dust measurement in mines in Japan. Evaluation of airborne dust in the working environment has been standardized in mass concentration since 1979. A device for dust measurement was developed to cope with the new measuring system and applied to work-environment measurement in representative mines of Japan. In consequence, the measurement control of airborne dust in the work environment should be established by the new measuring system.

INTRODUCTION

There were 32 coal mines, 10 lignite mines, 473 metal and nonmetal mines, 339 limestone mines, and 106 other mines in Japan in production in January 1981. The total annual production is roughly 270 million tons. The number of mine workers is about 66,000 and is gradually decreasing. The incidence of pneumoconiosis and suspected cases is 5.8% of all workers. The number of cases increased to 2.7 times that of the previous year because of a change in the pneumoconiosis diagnosis standard in 1978. The incidence follows that in the pottery manufacturing and foundry industries.

Accordingly, the Mine Safety Regulation was revised in 1978, and the requirements for airborne dust measurements and countermeasures were strengthened. The main points are:

1. education about dust;
2. expansion of the range and requirements for wearing dust respirators;
3. cleaning and removal of dust; and
4. measurement of respirable dust and improvement of the working environment.

MEASUREMENT OF AIRBORNE DUST

By 1949 a particulate number measuring system (konimeter) was in use. In about 1955 a relative concentration measuring system (mainly Tyndalloscope) was introduced for improvement of working environments. Since 1979 a measuring system based on gravimetric measuring instruments by filtering with a horizontal elutriator-type size classification apparatus has been adopted as the standard for the measurement of respirable dust, and is used together with the relative concentration measuring apparatus. The reasons why relative concentration is also used are:

1. Measurement of the change in airborne dust concentration is easy.
2. It is possible to measure dust at low concentrations.
3. Information at many measuring points is available.

However, it is necessary to calculate the conversion coefficient, because the results cannot be converted directly to mass concentration. The conversion coefficient depends on shape, diameter, concentration and color of airborne dust. Consequently, the conversion coefficient at each measuring point is required. In the new measuring system, the conversion coefficient at each measuring point is obtained by the simultaneous measurement with a horizontal elutriator-type low-volume sampler (HE-LVS, with suction air volume = 10 or 15 liter/min).

The indicated relative concentration can be converted to mass concentration by the conversion coefficient. Finally, the geometric mean concentration and the geometric standard deviation of all the measured results are calculated for the evaluation of dust concentration at each measuring point.

The relative concentration is measured continuously at 5–20 measuring points of each workplace within the work shift. The measurement is performed on two consecutive working days. It is required that the concentration be measured more than two times a year [Homma 1983].

CHARACTERISTICS OF THE HE-LVS

The setup of the HE-LVS used in mines in Japan is shown in Figure 1. The apparatus shown in Figure 1a is the widely used standard type to monitor environmental contaminant concentration, and is applied to collect particulates <10 μm. This apparatus collects respirable dust (under $\sqrt{50}$ = 7.07 μm) in working environments by adjusting suction air velocity in the elutriator. The apparatus in Figure 1b was developed by Tajiri [1977] for underground measurement, and is portable, with an installed battery. Its weight is about 6 kg and it is possible to collect dust continuously for 8 hours. The filter paper used for both the standard and portable models is 55 mm in diameter.

Sampling and measuring methods in this work were adopted in the Work-Environment Measurement Law. Measuring points were positioned at the center of each isometric division (usually ten divisions or more) in the workplace. All of the measuring points were equivalent. The measuring height was 1.5 m from the floor. The measurement was performed during work hours by sampling; where the work period is less than 30 minutes, the measurement was not carried out. Comparing the two types in underground workplaces, the results are approximately the same and have the same characteristics. The results of measurement in mines are shown in Table I.

The size distribution of airborne dust collected by HE-LVS is displayed in the Rosin-Rammler-Bennett diagram in Figure 2. (The airborne dust produced in mines has a Rosin-Rammler-Bennett distribution.) The collected sample was washed off in an alcohol solution by ultrasonic waves, and the extracted particle suspending solution is poured on a blood count cell. The size distribution is measured by image analyzer from a television

Table I. Respirable Dust Measurements (mg/m^3): Comparison of
Standard and Portable HE-LVS

Section of Mine	Apparatus	Shakanai Mine	Kosaka Mine
Crusher Room	Standard HE-LVS	1.65	6.11
	Portable HE-LVS-1	1.49	6.03
	Portable HE-LVS-3	1.46	6.10
Ore Bin Throw	Standard HE-LVS	1.45	1.89
	Portable HE-LVS-1	1.59	1.88
	Portable HE-LVS-2	1.78	1.84
	Portable HE-LVS-3	1.41	1.65

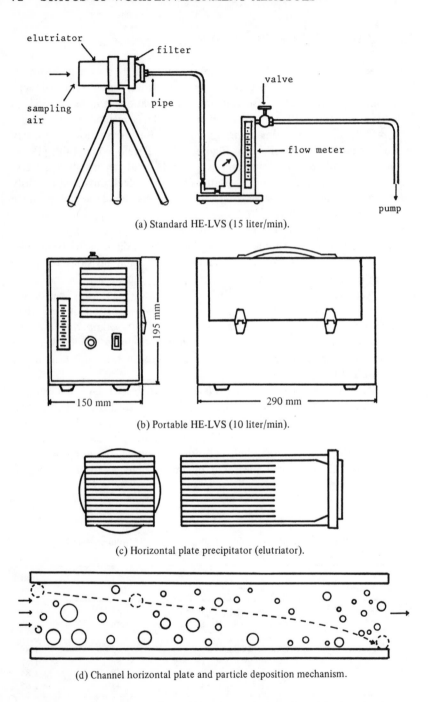

(a) Standard HE-LVS (15 liter/min).

(b) Portable HE-LVS (10 liter/min).

(c) Horizontal plate precipitator (elutriator).

(d) Channel horizontal plate and particle deposition mechanism.

Figure 1.   Construction of HE-LVS and the principle of particle size selection.

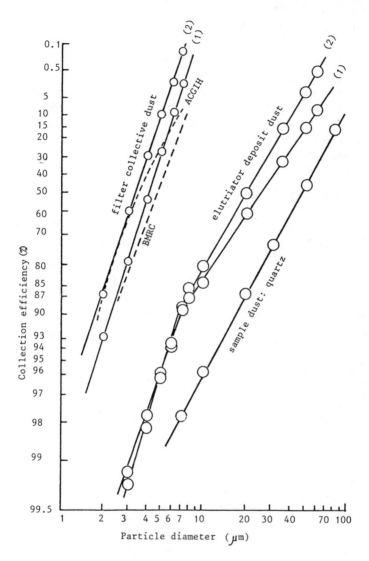

Figure 2.   Dust size distribution by HE-LVS.

screen image using 1500X magnification by an optical microscope. The size is displayed by Heywood diameter, and the weight percentage is calculated by computer [Tajiri et al. 1977,1978,1979].

In Figure 2, the dotted line expresses inspirable dust alveolar fraction published by American Conference of Government Industrial Hygienists

(ACGIH) and the British Medical Research Council (BMRC). It is shown that dust collected by filter and estimated to be respirable is under 7 μ. The collection efficiency for deposited dust sedimented in the elutriator deviates around 7 μ. This shows the selective collection.

Analysis of size distributions of the airborne dust collected in underground workplaces and of the reference particles are displayed in Tables II and III. The symbols n and d' in the tables express size distribution index and particle diameter calculated by size characteristics (R) at 36.8%, which are determined by the Rosin-Rammler-Bennett diagram.

DUST CONCENTRATION IN MINES

The geometric mean concentration and the geometric standard deviation of all the data measured by Tyndalloscope and digital dust indicator,

Table II. Size Distribution of Airborne Dust[a]

	Particle Size	
	n	d'
Alveolar Deposition Probability		
ACGIH, Nearly	2.3	4.2
BMRC, Nearly	2.3	5.8
Other, Nearly	2.2	3.0
Standard Testing Particle		
Latex Particle	2.1	1.8
Latex Particle	3.3	2.0
Fly Ash	2.9	7.5
Underground Workplaces		
Drilling		
23 Mines	0.5–2.5	1.0–9.0
Wet	1.2	4.2
Large Machine	2.8	3.8
Blasting		
After 30 min	0.8	8.3
Dry	3.0	6.6
Water Tamping	1.8	8.8
Coal Face Water Injection	2.4	7.2
Excavation		
Coal	0.5	9.8
Metal	1.5	7.0
Transportation		
Metal Ore	3.1	7.6
Mother Rock	3.1	11.0

[a]Rosin-Rammler-Bennett formula: $R = 100e^{-(d/d')^n}$, where R = aggregation weight (%), d = particle diameter (μm), d' = special character of particle diameter at R = 36.8% (μm), n = index of particle distribution.

Table III. Airborne Dust Characteristics by Drilling Workplace and Rock Kinds

Mine	Rock Kinds	n	d'
Ashio	Liparite	1.38	2.4
	Siliceous rock	0.92	4.0
	Acidic liparite	0.92	4.7
Hitachi	Siliceous-amphibolite-chlorite-schist	0.88	7.7
	Amphibolite-mica-schist	1.16	9.0
	Sericite schist	0.67	6.1
Ikuno	Andesite	0.66	6.0
Kawamori	Tuff	1.38	1.7
	Green tuff	0.92	4.5
Konomai	Liparite	0.88	7.2
Kushikino	Liparite	1.32	7.1
	Liparite	2.52	2.7
Osarizawa	Propylite	0.92	3.3
	Black shale	0.82	3.5
	Tuff	0.82	3.2
Shimokawa	Diabase	0.65	7.5
	Cu-bearing pyrrhotite	0.79	4.3
Takatama	Light-brown tuff	0.96	6.5
	Blackish tuff	0.88	3.8
	Tuff	1.80	1.0
Yanahara	Diabase	0.70	6.2
	Rhyolite	0.52	6.7
	Rhyolite	0.66	6.0

after calibration using the conversion coefficient from HE-LVS data, are shown in Table IV. Tyndalloscopes and digital dust indicators have been used widely in Japanese mines [Tajiri et al. 1980]. As shown in Table IV, it is possible to estimate the changing probability of respirable dust concentration at each measuring point, and the evaluation of dust concentration in the work environment has been improved by converting to mass concentration from the data of Tyndalloscope and digital dust indicators using the simultaneous measurement data of mass concentration meters and by examination of the geometric mean concentration and geometric standard deviation.

Although airborne dust had been estimated with relative concentration in the past, by adopting the new measuring method the environmental dust concentration in each workplace has been estimated with a time-weighted value of measured concentrations; the reliability of the measured values can be confirmed by calculating a geometric deviation. As the comparison of the respirable dust concentrations in different workplaces is made possible, the priority to carry out dust control in each workplace can be defined. Also, distribution of airborne dustflow in workplaces is made clear, and information about the working activity and the position of workers is obtained.

Table IV. Measured Results of Respirable Dust Concentrations
at Underground Exploitation Workplace

Mine	Geometric Mean Concentration (mg/m^3)	Geometric Standard Deviation (σg)
A:Ty[a]	1.18	1.75
Di	1.42	1.25
B:Ty	0.80	2.83
Di	1.08	3.00
C:Ty	1.33	2.02
Di	1.50	1.52
D:Ty	1.77	1.75
Di	1.71	1.27
E:Ty	1.16	4.18
Di	0.82	3.09

[a]Ty: Tyndalloscope, Di: digital dust indicator.

A relation between respirable dust and total airborne dust concentration must become clear in the future. The respirable dust concentration at 189 workplaces in mines in Japan is shown in Table V. It indicates that the countermeasures for dust control at workplaces should be strengthened and that the dust concentration is apt to decrease. However, the countermeasures are not enough at some workplaces; the Industrial Location and Environmental Protection Bureau has strengthened the promotive guidelines for dust control. The personal exposure concentration for respirable dust is being measured experimentally.

CONCLUSION

The system of airborne dust measurement at working environments in mines in Japan was introduced. The following points summarize the past investigation:

1. Airborne dust, especially respirable dust, has been measured since 1979.
2. Respirable dust is not measured sufficiently, but data are being accumulated.
3. Criteria for respirable dust at each dust producing workplace are becoming clear by concentration evaluation by the new measuring system for work environments. This contributes to the establishment of countermeasures to improve work environments.
4. HE-LVS has been useful as a selective classifying dust collector. According to present data, particles under 7.0 μ will invade the bronchus.
5. Personal exposure concentration is measured experimentally, but continuously.

Table V. Measured Results of Respirable Dust Concentration
in Main Work Environment in Japan

Location	Type of Mine	Kinds of Work	Number of Workplaces	Respirable Dust Concentration (mg/m^3)
Underground	Metal	Exploitation	25	4.05–0.25
		Excavation	10	3.45–0.20
		Transportation	8	1.87–0.13
		Crushing	15	1.85–0.27
		Welding	8	2.81–0.02
	Coal	Coal Face	5	45–2.50
		Excavation	4	2.80–0.80
		Rock Dusting	6	58–0.40
		Timbering	6	25–0.80
		Transportation	2	2.80–0.50
		Traffic	4	2.00–0.30
	Limestone	Excavation	2	10–0.30
		Transportation	2	7–0.20
		Crushing	18	7–0.25
		Other	6	3–0.16
Surface	Metal	Crushing plant (coarse)	6	0.18–0.02
		Crushing plant (fine)	3	0.35–0.04
		Ore bin (out)	2	2.80–0.03
		Ore bin (house)	2	3.15–0.12
		Machine tool plant	3	0.84–0.03
	Nonmetal	Crushing plant	7	6.57–0.35
		Dressing plant	5	2.35–0.03
		Transportation	10	5.38–0.02
	Limestone	Crushing	4	4.52–0.25
		Transportation	6	3.20–0.27
		Pelletizing	3	1.30–0.02
		Bagging	5	2.35–0.12
		Container Packing	3	1.34–0.03
	Refinery	Cu smelting	2	0.19–0.02
		Zn smelting	2	0.20–0.02
		Pb smelting	2	0.84–0.03
		Ni smelting	1	1.80–0.25
		Solvent crushing	1	1.85–0.35
		Pretreatment dryer	1	1.80–0.12

APPENDIX: MEASURING APPARATUS MANUFACTURERS

Digital dust indicator, standard HE-LVS:

Sibata Chemical App. Mfg. Co., Ltd.
3-1-25 Ikenohata, Taito-ku
Tokyo, Japan

Portable HE-LVS:

Dan Science Co., Ltd.
1-1-25 Owadamachi Hachioji-shi
Tokyo, Japan

Image analyzing system (Luzex 500):

Nihon Regulator Co., Ltd.
2-8-7 Kyobashi, Chuo-ku
Tokyo, Japan

REFERENCES

Homma, K. (1983) "Work-Environment Measurement Law and Its Present Status of Application in Japan," Chapter 4, this volume.

Tajiri, T. (1977) "Methods for Respirable Dust Measurement in Working Environments," Committee for Pneumoconiosis and Dust Counterplan, Mining Labor Accident Prevention Association, Japan.

Tajiri, T., M. Kohno and A. Omori (1977) "Results of Mine Working Environment Measurement by Portable Low-Volume Air Sampler," Committee for Pneumoconiosis and Dust Counterplan Annual Report.

Tajiri, T., M. Kohno and A. Omori (1978) "Results of Mine Working Environment Measurement by Portable Low-Volume Air Sampler," Committee for Pneumoconiosis and Dust Counterplan Annual Report.

Tajiri, T., M. Kohno and A. Omori (1979) "Results of Mine Working Environment Measurement by Portable Low-Volume Air Sampler," Committee for Pneumoconiosis and Dust Counterplan Annual Report.

Tajiri, T., M. Kohno, A. Omori and S. Kato (1980) "Measurement Results of Respirable Dust in Mine Working Environment and the Evaluation," Committee for Pneumoconiosis and Dust Counterplan Annual Report.

CHAPTER 6

CONTROL OF AEROSOL EXPOSURE IN THE SWEDISH WORK ENVIRONMENT

Staffan Krantz and Lars Olander

Aerosol Section and Ventilation Section
Research Department
National Board of Occupational Safety and Health
Stockholm, Sweden

ABSTRACT

An overview describing the state of occupational hygiene regulations, standards and measurement methods for aerosol exposure in the work environment is presented. The regulations are part of new legislation, the Work Environment Act, introduced in 1978. The aim of the legislation and the accompanying regulations is to eliminate any excess risk of injury or disease. The prime regulation for chemical hazards is the threshold limit value (TLV) list, which provides guidance for application of the Work Environment Act. In addition to the TLV list, there are a growing number of other regulations intended for specific toxic substances, such as quartz and asbestos. These regulations include, among other things, rules regarding air sampling practice and control measures. The strategy for exposure measurement has so far included two steps. First, a detailed investigation must be carried out. If this shows that the levels are below the TLV in question, a program of yearly control measurements is to be established. If, however, the TLV is exceeded, control measures must be taken. After this the full measuring program must be repeated. Since the 1960s, all aerosol exposure measurements for checking compliance with the TLV are carried out by personal sampling. It is specified in the TLV list whether total and/or respirable dust is to be sampled and analyzed.

INTRODUCTION

In 1978 a new Work Environment Act came into force in Sweden. This act superseded a previous Workers' Protection Act. The rules issued by the National Board of Occupational Safety and Health (NBOSH) under the Work Environment Act are called ordinances. Rules issued under the previous Workers' Protection Act are called directions, and will apply until they are replaced by rules issued under the new legislation.

Ordinances and directions are of several different types. For air contaminants, the most important is the new TLV list that comes into force in 1982. Other rules for air contaminants also exist. The following ones apply to aerosols:

1. material-specific rules, e.g., for asbestos and cadmium;
2. process-specific rules, e.g., for paint spraying and mining,
3. equipment-specific rules, e.g., for personal safety equipment; and
4. general rules, e.g., for premises and measures against air contaminants.

AEROSOL REGULATIONS

The TLV list contains permissible concentrations for about 100 different substances, and the levels are expressed as an 8-hour time-weighted average (TWA) value, a ceiling value or a short-time mean value. (Sometimes two of these are given.) Most of the levels are approximately the same as or somewhat lower than the levels in the TLV list adopted by American Conference of Governmental Industrial Hygienists for 1980. In this ordinance there are also lists of substances that are carcinogenic, skin-absorbing or irritating. The contents of the other types of ordinances and directions regarding aerosols are discussed here using asbestos, mining, personal safety equipment and measures against air contaminants as examples.

The directions for asbestos rule out the use of asbestos in general, but there are exceptions. When working at places where the presence of asbestos is suspected, there are rules regarding control measures, personal safety equipment, and measurement and analytical methods. For other fibers and silica dust there are also rules regarding measurement and analytical methods.

The mining and quarrying directions contain rules for designing ventilation systems taking into consideration radon, diesel exhausts, silica dust and gases from blasting. There are also directions for when and how to measure the air contaminants.

The directions for personal safety equipment are divided into several parts. Two of these deal with respirators. The first one contains general requirements for respirators and discusses how to make leakage and collection tests, and which type of respirator to use when working with different aerosols (and gases). The other one tells how to use and maintain different types of respirators.

The ordinance for measures against air contaminants is a little different from the abovementioned directions. It contains no definite instructions on how and when to apply different measures against air contaminants. It is only a list of suggestions, e.g., replacement or alteration of a substance or process; confinement of the work, the process or the worker; process ventilation; and restriction of work to special hours.

THRESHOLD LIMIT VALUES

In the TLV list there are approximately 50 different substances that can exist as aerosols. Most of them have only an 8-hr TWA value expressed in mg/m^3 or fiber/ml. For some 10 different dusts and fumes, the new TLV list also has a TWA value for the respirable part. A few aerosols also have a ceiling value or a short-term value, either for the total dust or for the respirable part. For iron oxide, silica dusts and zinc chloride, the TWA are given only for the respirable part. Respirable dust is defined according to the Johannesburg Convention.

Since the 1960s, all aerosol exposure measurements for checking compliance with the TLV have been carried out by personal sampling. As mentioned above, it is specified in the TLV list whether total dust and/or respirable dust is to be sampled and analyzed. The method in use is the membrane filter method, where the filter is placed in a 37-mm cassette. Sampling is carried out with the open-face cassette for total dust and asbestos fibers. When sampling respirable dust, the filter is placed in a cassette within a cyclone presampler, which fulfills the Johannesburg criteria for respirable dust. Besides the filter, cassette and, in the respirable dust case, cyclone, the sampling equipment consists of a battery-operated pump, an external rotameter and a carrying belt. The filter cassette is placed in the "breathing zone" at the worker's shoulder, and the pump hangs from the belt. After sampling, the cassette is mailed to a laboratory for evaluation of the contaminant and the concentration.

The strategy for exposure measurements up to 1981 included two steps. A detailed investigation (including, for silica dust, total and respirable dust) must be carried out. If this investigation shows that the levels are below the actual TLV, a program of yearly control measurements is to be established. If, however, the TLV is exceeded, control measures must be taken. After this, the full measurement program must be repeated.

CONTROL MEASUREMENTS

This strategy is, in principle, still valid, but for silica dust some changes will be introduced in 1982. In a new quartz ordinance it is stipulated that if quartz-containing dust is generated by an operation, the employer must

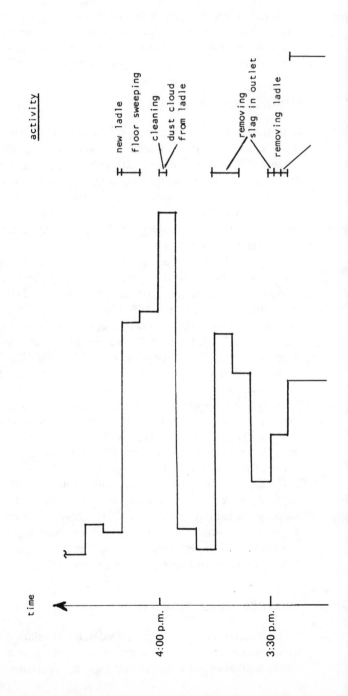

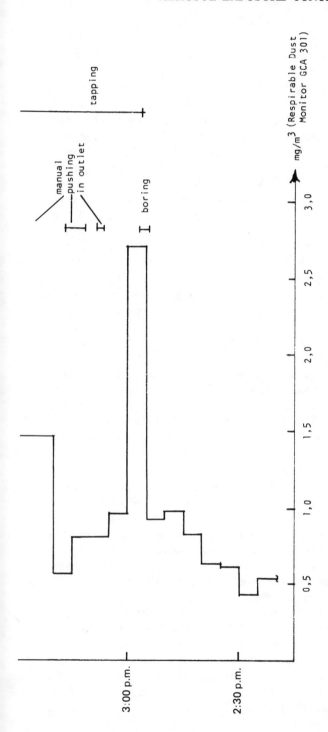

Figure 1.   Dust concentrations in a steel plant near a blast furnace.

perform a dust measuring survey. This must take place if it is not obvious, due to the operation in question, that the TLV is not exceeded, which actually means a level of 10% of the TLV. Examples of such operations are handling of quartz-containing materials at, for example, a dental laboratory, where only grams are handled at each occasion, or an enclosed feeding operation at a chemical plant. Also, it is not necessary to carry out an exposure measurement at a temporary operation where dust is generated from equipment, supplied with a local exhaust, provided an earlier measurement at a similar operation has proved that the equipment does not generate dust concentrations that exceed the TLV. If a measurement is found necessary, it should be carried out within six months after the operation was started or after significant changes have taken place. After this, measurements shall be carried out once a year, as has been the case earlier. If, however, and this is new, the concentrations do not exceed 20% of the TLV, no more measurement need take place. If the concentrations exceed 20% of the TLV, but are lower than 50% of the TLV, the local inspectorate may allow for intervals longer than a year between the measurements. If the measurement shows that the TLV is exceeded, a new measurement must take place within six months.

The measurement must include respirable quartz and, if high temperatures are present in the process, cristobalite and tridymit. The sampling must cover one whole day of normal production. If the exposure only takes place during a shorter period, this period shall be sampled. If the work is carried out as a shift work, measurements shall be made at the different shifts, provided the work differs. The measurement shall cover the exposure for a number of workers sufficient to judge the exposure for the total workforce. The personnel who carry out the sampling are educated for this. Courses are regularly arranged for this purpose.

In addition to the first measurement at a plant, it is recommended that an investigation be made to find out the significance of different emission sources to the total dust concentration and the spreading of airborne dust in the plant. By doing this, better knowledge is achieved for implementing a suitable control program. Such a dust source investigation can preferably be carried out by different direct-reading instruments. An example of results from such an investigation in a steel plant is shown in Figure 1.

It is, of course, of great importance that dust-suppressing methods and equipment be controlled and maintained regularly, and that the measurements give a representative picture of the normal dust conditions. As a result of the regulations for silica dust in Sweden and the measurements and measures that thereby have been carried out, dust concentrations in many dusty operations have been cut down. An example of this is the situation in iron foundries; Figure 2 shows the situation 10–12 years ago for various operations compared with the situation 2–3 years ago.

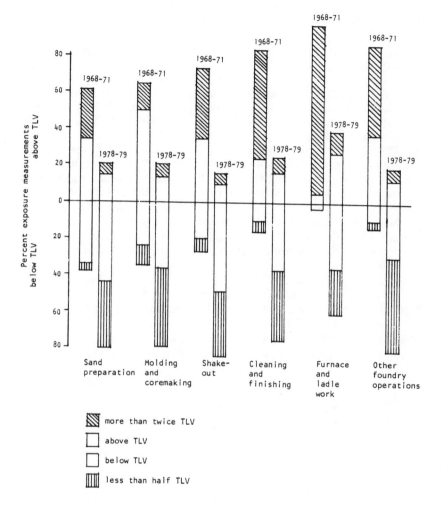

Figure 2.   Dust concentrations in iron foundries, 1968–1971 and 1978–1979.

Due to the importance of silicosis as an occupational disease, special emphasis has for many years been given to regulations in this area, including measurements. There are, however, regulations for other aerosols (e.g., asbestos and chromic acid mists) that also demand measurements. For the moment a special "measuring ordinance" is being developed which will recommend principles and rules for measurements of all substances which occur on the Swedish TLV list.

BIBLIOGRAPHY

Holm, L. (1981) "The Fight Against Silicosis," in *Annual Report 1980* (Stockholm, Sweden: National Board of Occupational Safety and Health).

Jansson, A. (1977) "An Airborne Dust Concentration Instrument That Employs Absorption of Beta Radiation—GCA Respirable Dust Monitor," National Board of Occupational Safety and Health Investigation Report, Stockholm, Sweden.

NBOSH (1967) "Personal Safety Equipment: Respirators," Direction 45, National Board of Occupational Safety and Health, Stockholm, Sweden.

NBOSH (1974) "Mining and Quarrying Directions," Direction 67, National Board of Occupational Safety and Health, Stockholm, Sweden.

NBOSH (1975a) "Asbestos," Direction 52, National Board of Occupational Safety and Health, Stockholm, Sweden.

NBOSH (1975b) "Personal Safety Equipment: Use of Respirators," Direction 45, National Board of Occupational Safety and Health, Stockholm, Sweden.

NBOSH (1980) "Measures Against Air Contaminants," Ordinance 11, National Board of Occupational Safety and Health, Stockholm, Sweden.

NBOSH (1981a) "Quartz," Ordinance 16, National Board of Occupational Safety and Health, Stockholm, Sweden.

NBOSH (1981b) "Threshold Limit Values for Air Contaminants," Ordinance 8, National Board of Occupational Safety and Health, Stockholm, Sweden.

"Work Environment Act" (1977) Swedish Cod of Statutes 1160.

CHAPTER 7

WORK-ENVIRONMENT AEROSOL RESEARCH IN FRANCE: A REVIEW

M. M. Benarie

Départment Pollution des Atmosphères
Centre de Recherche
Institut National de Recherche Chimique Appliquée
Vert le Petit, France

About 90–100 work groups, laboratories, schools and companies are on record of the Ministry of the Environment as dealing or having dealt with air pollution–related topics, which, more often than not, includes some research about particulates. The interest of most of these in aerosols is either marginal, occasional or limited to some very special application. Six laboratories have sustained tradition, interest, trained manpower and a sizable amount of publications in the aerosol field. These are:

1. Paris University, formerly Paris VI, under the management of J. Bricard, well known for his studies in electrical properties of aerosols, nucleation, small particles, etc. (aerosol research is now carried on by A. Renoux at Paris XII);
2. Laboratoire National d'Essais, which does surveys, measurements, and standardization (not research-oriented);
3. CERCHAR, the research institute of the national coal industry which deals with all particulate problems related to coal, beginning with the respiratory hazards in the mine, through dust problems in handling and shipping coal, to particulate abatement in coal combustion.
4. CETIAT, a laboratory maintained by the manufacturers of heating and ventilating equipment, oriented toward this professional branch; they test and certify all kinds of dust abating machinery and its installation, conduct research to reduce the particulate emissions of various burners; and also work on particulate abatement at high temperature;
5. CEA (Atomic Energy Commission), which has a strong team specializing in dust problems of uranium mines, and does a lot of research on plutonium and sodium aerosols and basic research concerning nucleation and formation of very fine aerosols, mainly in cooperation with Paris University;
6. Institut National de Recherche Chimque Applique (IRCHA).

This short list shows that five out of the six leading laboratories are busy mainly outside our specific topic, aerosol research in the work environment, leaving us to concentrate on IRCHA. The French acronym stands for National Institute for Chemical Research. IRCHA, oriented toward contract research, is controlled and partly financed by the Ministry of Industry. Its aerosol laboratory is headed by J. C. Guichard.

The traditions of work-environment aerosol research within the IRCHA reach back to the early 1950s, when IRCHA designed, built, standardized and used dust measuring (counting and weighing) apparatus for industrial environments. Many early dust generators were developed at that time, along with measurement methods. The first international meeting to discuss these matters was held in 1952, and the 15th in May 1982.

During the 1960s, activity concentrated on microscopic dust measurement techniques, which have been consolidated and standardized on behalf of the steel and coal industries of the European community. At the same time, industrial dust respirators have been developed further, tested and standardized. A great number of industrial filters of all kinds—air conditioning, automotive, high-efficiency—have been certified by novel methods; filtration theory developed, and testing methods improved. Research in aerosol physics conducted in the laboratory during this period was mainly aimed at these targets. A few typical papers of this time are: Avy et al. [1963,1967], Benarie [1963,1964,1966,1969], Benarie and Bodin [1968] and Benarie and Panof [1970].

From the 1970s on, the idea of "work environment" has been extended beyond the respiratory hazards to the industrial product itself, into cleanroom technology. Patents covering improved laminar flow hoods mark this stage.

At the present, the attention of the aerosol laboratory is oriented toward research and development of:

1. aerosol-generating methods for better simulation ("Puldoulit") using fluidized beds;
2. aerosol dilution chambers as complement of the generators; and
3. the "Jetimeter" [Guichard et al. 1980], for continuous measurement of particulate concentrations between 0.001 and 1 mg/m^3.

The above three devices are patented and are built by the IRCHA and commercially available from it.

A further domain of activity is the development of a dielectrophoretic particulate filter. A strongly heterogeneous electric field is maintained over a fixed bed of glass spheres in the millimeter size range. The investigations concern the energetics and the possible improvement of the filtration efficiency of the bed, as compared to a bed devoid of electric field. Other activities of the aerosol laboratory, which are not within the work-environment field, are not mentioned here.

REFERENCES

Avy, A., M. Benarie, K. H. Schmitt and A. Winkel (1963) "Vergleichende Untersuchungen an Staubmessgeräten in einer Staubkugel und in einer Sinteranlage," *Staub* 23:1.

Avy, A., M. Benarie and F. Hartogensis (1967) "Comparison of Dust Sampling Methods and Instruments," *Staub Engl.* (November 1967), p. 1.

Benarie, M. (1963) "Der Einfluss der Form Fester Staubteilchen auf Ihre Filtrierbarkeit," *Staub* 23:50.

Benarie, M. (1964) "Ein Eichmass für Mikroskopische Sichtbarkeitsgrenze und das Auzählen von Staubteilchen," *Staub* 24:514.

Benarie, M. (1966) "Eine neue Mikroskopische Korngrössenmessmethode," *Staub* 26:207.

Benarie, M. (1969) "Einfluss der Porenstruktur auf den Abscheidegrad in Faserfilter," *Staub* 29:74.

Benarie, M., and D. Bodin (1968) "Contrôle des Appareils de Protection Individuelle Contre les Gaz et les Vapeurs," Cahier Notes Doc. INRS 602-52-68.

Benarie, M., and S. Panof (1970) "Notes dur le prélèvement des Particules Solides en Suspension dans un Gaz à Écoulement Turbulent," *J. Aerosol Sci.* 1:21.

Guichard, J.-C., A. Gaillard and M. Lamauve (1980) "Some Applications of the 'Jetimeter'," *Atmospheric Pollution; Proceeding of the 14th International Colloquium*, M. Benarie, Ed. (Amsterdam: Elsevier), p. 339.

CHAPTER 8

REVIEW OF FINNISH RESEARCH ON WORK-ENVIRONMENT AEROSOLS

T. Raunemaa

Department of Physics
University of Helsinki
Helsinki, Finland

ABSTRACT

A review of current research on aerosols in Finland shows that about 25% is devoted to basic research, while 75% is directed toward applications. Research carried out in government institutes and universities on work-environment aerosols is summarized, and the contribution to aerosol research derived therefrom is evaluated. Lack of measuring instruments is the main reason for slow progress in some study areas. An overview of administration and legislation in the work environment field is presented along with a comparison of the exposure limits for some compounds in air in effect in Finland.

GENERAL

Systematic research on environmental aerosols, particularly in work environments, is a relatively new field in Finland—less than ten years old. Earlier research in this field concentrated mainly on gaseous impurities and particles of large size, the more hazardous small sizes and particle transformations being outside the available research capabilities. Recent years have seen an apparent change in attitude as research interest has centered more and more on smaller particle sizes and difference measurements. However, a lack of suitable measuring instruments has restricted progress in these aspects of research.

The basic research being carried out in six universities and three technical universities (Figure 1) derives from the individual interests of the research groups and persons concerned, as no instruction, or only very little, has been given in the aerosol field in Finland. In aerosol science, no regular instruction on the higher academic levels has been given; studies in meteorology and ventilation techniques have helped this situation somewhat. Nevertheless, despite of the newness of the research area and the perhaps poor educational background, research on work-environment aerosols has produced some interesting results and promising technical advancements.

Most of the work in the field is carried out through application of commercially available measuring instruments to monitor the present state of various work environments, there being very little involvement in pure basic research. This work has been carried out principally by the Institute of Occupational Health (IOH), a government-financed central institute with six regional units. This research activity is focused on the problems of work and health typical of Finland. The information necessary for drafting occupational health legislation is also gathered by IOH.

Theoretical aerosol research is primarily directed at developing new instruments, and almost not at all at investigating the processes that are involved in aerosol transformation. The transport problems are considered by the people working in meteorological institutes. In any approach toward studying the fine-particle aerosols <0.1 μm in size, the resources available are very scarce. Any broader systematic research in this part of the field is thus inhibited. At the first Aerosol Symposium on Fine Particle Aerosols [Academy of Finland 1981], principally all those currently working in the aerosol field were gathered together (80 persons attended the symposium). Thus, evaluation of the symposium reports may give greater insight into the present status of the field. The presentations at the symposium can be divided into two groups, basic and applied research, as follows:

1. basic research (25% of the reports); and
2. applied research (75% of the reports): (1) total number or mass determinations (31%), (2) size determinations (13%), (3) composition and concentration determinations (25%), and (4) other properties, e.g. radioactivity (6%).

The work environment was under study in 25% of the reports.

The general aspects of the Finnish aerosol research can be summarized as follows:

1. Stable properties (size, form, mass, number, composition, radioactivity, etc.) are studied by internationally standard methods or by proven sophisticated methods with acceptable accuracy. Only fine particles are left outside this category.
2. Dynamic properties (i.e., transport and transformation processes) are seldom investigated. Transport of particles is considered mainly by model calculations and measured usually only by cumulative, slow methods. Improvement of continuous detection methods with fast-time resolution is much desired.

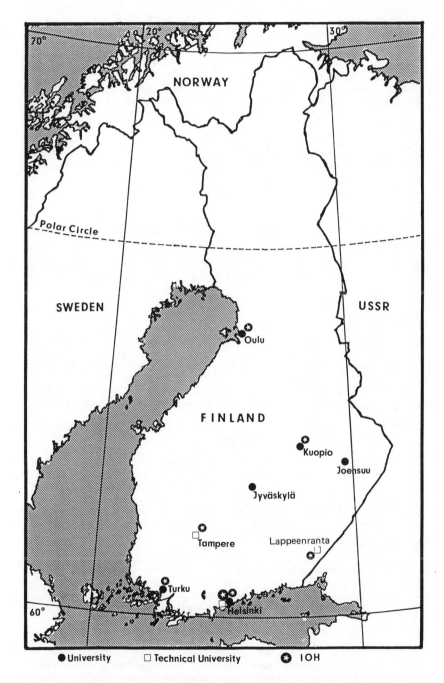

Figure 1.   Universities, technical universities and IOH laboratories in Finland working in the aerosol field.

3. Studies on accumulated effects are numerous; thus, monitoring surveys have been carried out in the iron and steel industry and in the asbestos industry. Also, nonuranium mines have been studied for radioactivity levels.
4. The structural details of the aerosols are either never or only very seldom investigated. Surface effects are thus incompletely known.
5. An intensified consideration of humidity, strong influence of optical properties of particles and alterations to the results caused by the choice of sampling line are all prominent in present aerosol research.

CURRENT RESEARCH IN FINLAND

Applied Research

IOH focuses the main part of its work on the problems typical of Finland. As the following survey is directed at those problems, very little attention will be paid here to some otherwise important projects. A list covering all research projects current in 1981 can be found in the official booklet by IOH [1982].

1. Silicosis in Finland, 1965–1977, occurrence and mortality, e.g., pulmonary and lung function testing: in 1972–1978, 3500 workers were studied for quartz dust and partly for polyaromatic hydrocarbons (PAH). Dust sampling was carried out for concentration determinations. The quartz limit of 0.20 mg/m^3 was exceeded in ~50% of the foundries. PAH were determined by gas chromatography.
2. Asbestosis: asbestosis, where mortality due to pulmonary cancer is ~35% in Finland, is presently studied by neurological examination of a study group (120 persons).
3. Ferrous metal aerosols from welding and grinding of steel in foundries, shipyards and steel manufacturing: analyses are performed of the chemical [by neutron activation analysis (NAA) and atomic absorption spectroscopy (AAS)] and physical properties of airborne dust. Inhalation studies of retention and clearance are carried out using rats. The magnetic method is used to measure the retained amounts of dust in the lungs; chromium and nickel (from aerosols) in urine are determined. Strongly magnetic magnetite is detected by remanence magnetism. Retained dust is deduced by its relative amount. The detection limit is 20 μg of magnetite in the lung. Transfer in the lung is studied by relaxation effects.
4. Allergy caused by feed yeast: emissions of feed yeast from a single-cell protein plant can cause respiratory allergy. The implications for sensitive persons will be studied.
5. Chromium-lignin, its kinetics and monitoring of chromium lignosulfonate produced in a sulfite pulp mill: chromium-lignin containing ~6% Cr and <1% Fe appears as a very fine dust in concentrations of 0.002–0.04 mg/m^3. Cr(III) lignin seems to differ from other Cr(III) compounds. Toxicology and kinetics in the body are evaluated using air sampling, urine testing and blood sampling.
6. PAH in iron foundries: by personal sampling of air (glass fiber filters) and by urine samples, the PAH components are determined by gas chromatography, and mutagenic activities are measured. An exposure estimation project on PAH will be carried out. A change in the foundry techniques will remove this PAH hazard.
7. Airborne contaminants in the production of pulp are measured.

8. Peat dust in peat bog workings is studied.
9. Spore dust in agriculture, e.g., in molding hay: allergic alveolitis (farmer's lung disease) may be caused by exposure to mold and actinomycete spores, which become airborne in the handling of hay, grain, bedding or straw. Molding is caused by wet storage. Agents such as bis-ammoniumpropionate are used in molding. Andersen sampling was performed in 1979–1980 on two farms. Spore count fluctuated, with a clear increase from ~30,000/m^3 (July) to 1–2 million/m^3 (October). Mesophilic molds were the most numerous (*Aspergillus, Penicillium, Clodosporium mucor*). *Aspergillus umbrosus* was present at levels of ~65% in the airborne dust. Amounts of thermotolerant and thermophilic molds were increased in hay stored wet.

Basic Research

Basic research with aerosols is carried out in universities and research institutes. In universities, optical detection methods are in use at Joensuu University and Lappeenranta Technical University. Magnetic detection of metals and metal fumes are studied at Oulu University and in Helsinki (IOH). Aerosol monitoring instruments and ventilation effects are investigated at Tampere Technical University, and transport of ions at Jyväskylä University. The aerosol study group is distributed between Kuopio, Turku and Jyväskylä Universities. Light radar developments are carried out at Helsinki Technical University and transport mechanisms, impaction processes and elemental analyses are researched at Helsinki University.

The following survey is directed to these basic research projects and gives a short outline of the work performed. The basic research is financed principally by the Academy of Finland (National Research Organization) and by the individual universities. Some of the groups have active contacts with the research carried out abroad, but very little funding is involved in this cooperation. In many cases, universities and research institutes have close national cooperation. The number of academic persons performing basic research is about 30. In most cases, only part of these individuals' research interests can be devoted to aerosol research.

Production of aerosols:

- dry dust (re)dispersion by using mechanical rotation (or vibration) and a fluidized bed; concentrations up to 200 mg/m^3 can be achieved;
- solid-liquid aerosol generator based on the use of a standing (ultrasonic) wave; about 200 kPa pressure with 250 kHz applied in the test device;
- droplet growth/evaporation (heterogeneous nucleation and condensation) around ~1 μm size at different relative humidity conditions; model calculations will be included; and
- automated apparatus for the production of welding fumes.

Transport of aerosols:

- electrical aerosol monitor: detection of 0.01- to 1-μm particles by diffusional charging and continuous operation principle; system sensitivity depends on the particle diameter;
- lidar detection: dynamic range found problematic in detection;

- optical extinction technique using three different wavelengths in situ: computer programs developed for Mie theory calculations (effects of particle population changes on the extinction are being studied); and
- aerosol elemental analysis by the proton-induced X-ray emission method with analysis from concentrated samples.

Properties of aerosols:

- the oxidation states of metals are studied by X-ray photon spectroscopy: chemical shift is used for the separation of different oxidation states;
- radioactivity within the particles; attachment to and transport into the environment; and
- microorganisms in the agricultural environment: levels of different molds determined.

Effects and detection of aerosols:

- magnetic detection of inhaled metal aerosols: magnetic properties of ferrous aerosols are used for in vivo measurements; remanence and other system parameters are calibrated under laboratory conditions (the limiting value of dust sample amount is ~1 mg); and
- origin and effects of particles in ventilated environments: effects of outdoor to indoor transport and settling on the surfaces are studied; effects of the cleaning process and surface material on the particle concentration in air are considered.

LEGISLATION OF WORKING CONDITIONS

Administration

The main responsibility in labor protection is borne by the Ministry of Social Affairs and Health. This includes legislation, orders and budgeting. The ministry also issues guidelines to the expert body, the National Board of Labor Protection, and to its districts. The district authority extends to all places of work where hazardous substances or methods are used. A Communal Health Board takes care of small working places with small assumed health risk. In 1978 IOH was made subordinate to the Ministry of Social Affairs and Health. IOH has six regional untis in Helsinki, Tampere, Turku, Oulu, Kuopio and Lappeenranta. The central institute is located in Helsinki. The research carried out by IOH is used for the development of legislation. The Labor Protection Administration and its subordinates cooperate with numerous international groups.

Regulation

In Finland, the Labor Protection Act (1958) regulates the working conditions in general, the Mining Act being applicable to miners' work environment. Supplementary government orders and instructions regulate the working environment more closely. The regulations concerning the work environment air include listing of dangerous chemicals. As the maximum working hours in Finland are 40 per week, in some plants averaging 36 a week, an 8-hr period is usually used for the daily exposure measure. Regulation of hazardous substances in work-environment air also include government orders and instructions for standardized measurements.

Government orders give detailed instructions for measurements, such as the measurement of dust concentration in workplace air by a filter method $(0.1-100 \text{ mg/m}^3)$; relative humidity is restricted to $<90\%$ and temperatures to $<50°C$, the measurement of formaldehyde by the chromotropic acid method $(0.1-1 \text{ mg/1})$, and the measurement of organic vapors by the charcoal tube method $(\sim 1-10,000 \text{ mg/m}^3)$. The exposure limits for selected elements and compounds are listed in Table I.

APPENDIX: MAIN GOVERNMENTAL RESEARCH INSTITUTES IN THE FIELD OF AEROSOL RESEARCH

Institute of Occupational Health

Since 1978 this has been under the Ministry of Social Affairs and Health. On January 1, 1981, there were about 300 staff members in the central institute and the six regional units around Finland. Scientific research consists of about 35% of total work. On January 1, 1981, about 50 persons were

Table I. The Exposure Limits for Some Compounds in Finland
[National Board of Labor Protection 1981]

	8 hr	15 min
Arsenic	0.1 mg/m^3	
Asbestos[a]	2 fiber/cm^3	
Acetone	500 ppm	625 ppm
Dichloromethane	100 ppm	250 ppm
Epichlorhydrin	0.5 ppm	
Inorganic Dust	10 mg/m^3	
Formaldehyde		1 ppm
Kristobalite[b]	0.1 mg/m^3	
Organic Vapors, PAH	50 ppm	75 ppm
Lead	0.1 mg/m^3	
Nickel, Metallic	1 mg/m^3	
Nickel, Compounds	0.1 mg/m^3	
Sulfur Dioxide	2 ppm	5 ppm
Styrene	50 ppm	100 ppm
Tetrahydrofurane	100 ppm	150 ppm
Trichlorethylene	30 ppm	45 ppm

[a]Length >5 μm.
[b]Length <5 μm.

working on aerosol research and related subjects. IOH carries out routine analysis using membrane filters. Annual analyses include 2000 dust samples, 500 quartz determinations, 500 metal determinations, and 200 asbestos determinations. Also, personal protection devices, electric filters, and PAH determination methods (e.g., for benzo[a]pyrene) are tested. Only a very few animal tests are carried out. Some special facilities are available: aerosol generators, classifiers, optical sizing devices, concentration analysis instruments, scanning electron microscopy with X-ray spectrometer and magnetometer for metal analysis.

Institute of Radiation Protection (IRP)

IRP is under the Ministry of Social Affairs and Health. Its research laboratory analyzes radioactivity found in air. This radioactivity is due to radon and radon daughters in soil and water, nuclear power stations, and other manmade sources. Detection of radon is carried out by ZnS detectors, and radon daughters are studied by integrative methods, e.g., cellulose nitrate films. About 10 persons are currently involved in related aerosol studies.

Technical Research Center of Finland, Ministry of Commerce and Industry

A dust laboratory was established in 1980 to concentrate on aerosol instrument testing. The Technical Research Center performs emission measurements on stack gases in industry and energy production. These emission measurements are carried out on a continuing basis.

Finnish Meteorological Institute, Ministry for Traffic

The institute concentrates on all kinds of ambient air determinations. Much work is performed on mathematical modeling of gas dispersion. Control methods are also tested by the institute.

REFERENCES

Academy of Finland (1981) *Proceedings of the Symposium on Fine Particle Aerosols, March 13-14, 1981, Lammi, Finland* (Helsinki: The Academy of Finland).

IOH (1981) "Current Research Projects 1981," Institute of Occupational Health, Helsinki, Finland.

National Board of Labor Protection (1981) "Chemical Substances in Workplace Air," Tampere, Finland.

CHAPTER 9

BUREAU OF MINES RESPIRABLE DUST MEASUREMENT AND MONITORING PROGRAM

K. L. Williams and G. H. Schnakenberg, Jr.

Pittsburgh Research Center
Bureau of Mines
U.S. Department of the Interior
Pittsburgh, Pennsylvania

ABSTRACT

This chapter will review current projects and discuss the future direction of the program. Projects presently underway include the development of a mining machine–mounted respirable dust monitor based on light scattering; the development of direct-indicating personal dust exposure monitors, one using light scattering and the other using a tapered element oscillating microbalance technique; a feasibility study of a light reflectance technique for measuring the mass of dust deposited on a filter; an in-house evaluation of various existing dust monitors; and two projects related to the determination of the quartz content of dust samples. The technical aspects of the various developments will not be discussed in detail; rather, the philosophy and rationale behind the development or investigation will be explained. Also, the chapter will point out the need for and nature of a planned respirable coal mine dust monitoring strategy study.

INTRODUCTION

This chapter discusses the Bureau of Mines (BOM) respirable mine dust monitoring and measurements program. Although some technical discussion is included, the chapter is meant primarily to present and discuss the rationale

99

behind the program. Most of the research and development effort is directed toward monitoring respirable coal mine dust, because of the great number of coal mining operations and because of the well known dangers of coal-worker's pneumoconiosis. Often, however, technology or methodology used in coal mines can be applied to other dust monitoring situations as well.

Coal mine dust personal sampling units (CMDPSU) have long been used to determine mineworker exposure to airborne respirable coal mine dust. The CMDPSU, often referred to simply as a "personal sampler," is a gravimetric device that can be worn by the worker or placed in the vicinity of work. A small battery-operated air pump draws dusty air through a dust precollector and filter cassette combination. The precollector, a 10-mm-diameter Dorr-Oliver nylon cyclone, captures large particles, such as those that deposit in the upper respiratory passages. Small particles, such as those that penetrate to the lower respiratory passages, pass on to a preweighed filter. After sampling is completed, the filter is reweighed. Knowing the mass of dust collected, the sampling flowrate, and the total sampling time, one can calculate the average airborne respirable dust concentration over the sampling period. Since health effects are not believed to be related to short-term excursion levels of dust concentration, the CMDPSU gravimetric approach served well to enforce the 2-mg/m^3 mass concentration standard for personal exposure and to gather epidemiological data. However, since sampling is usually performed over an entire 8-hr shift, and since days or even weeks may pass before dust level information is made available to the workers, emphasis has not been placed on using the CMDPSU to monitor dust for control purposes. However, since fast response technology for the measurement of airborne respirable mine dust is available today, BOM is taking a much closer look at the possibilities of monitoring for the purpose of dust control.

RESEARCH PROGRAM

The respirable dust section of the Instrumentation Group of Electrical Safety and Communications at the Pittsburgh Research Center conducts dust instrumentation and measurement methodology research, either in-house or under contract to the private sector. Presently, the work we perform generally falls into one of three categories:

1. direct requests from the Mine Safety and Health Administration (MSHA), U.S. Department of Labor;
2. feasibility studies of measurement technologies as applied to the mining industry;
3. development of hardware where a need is apparent or application appears very likely.

The first category needs little discussion. Occasionally in the conduct of its mission to enforce the 2-mg/m^3 respirable dust standard, MSHA identifies a specific monitoring problem or need. Once MSHA communicates that need to BOM, it makes every reasonable effort to develop new instrumentation or suggest existing instrumentation that will meet MSHA needs.

Feasibility studies may take one of two forms. BOM researchers may investigate a new or existing measurement technology to determine whether or not it can be used to sense airborne respirable mine dust, primarily coal dust. Once a technology has been identified, BOM determines if the technology is adequate for the intended monitoring purpose, if it can be incorporated into a device that would withstand the rigors of the mining environment, if it can be made intrinsically safe for operation in explosive methane atmospheres, etc. Feasibility studies might also involve evaluating existing aerosol monitors designed for more general monitoring applications. The objective of these evaluations is, of course, to see if the instruments could be used in the mining environment. The final category is self-explanatory. Hardware development takes place when fairly well defined specifications for the device can be cited.

Projects currently underway include the development of a mining machine-mounted respirable dust monitor; development of direct-indicating personal dust exposure monitors using two distinctly different technologies; a feasibility study of a light reflectance technique for measuring the mass of dust deposited on a filter; an evaluation of various existing dust monitors; and two projects related to the determination of the quartz content of dust samples.

MACHINE-MOUNTED RESPIRABLE DUST MONITOR (MMRDM)

The concept of putting a monitor on a mining machine to measure respirable coal mine dust has been discussed for a number of years. Strong recent interest by MSHA, however, has brought about a concerted effort to produce hardware. The technical details of the device are discussed by Lilienfeld [1983]. Briefly, however, MSHA requested a device that would be mounted on and obtain power from the mining machine. The device under development uses light scattering to measure airborne respirable coal mine dust. Both instantaneous and shift average dust concentrations can be displayed to the mining machine operator. The shift average is defined by:

$$\text{shift average} = \int_0^t \frac{C(t)dt}{480}$$

where   C(t) = instantaneous dust concentration
             t = sampling time (min)

The constant 480 is in minutes and corresponds to a full 8-hr shift. Shift averages will be stored in memory for later retrieval. The device will incorporate a filter system to make an average gravimetric measurement to verify average light scattering measurements or to facilitate calibration of the light scattering sensor. MSHA has also requested the MMRDM have the ability to shut off the production function (the cutting head in the case of a continuous miner) of the mining machine if the shift average for a particular shift exceeds a predetermined level. Hardware is being developed to meet these specifications.

BOM sees this MSHA-requested development as a great opportunity to investigate the merits of monitoring for control. Mining under the present monitoring program has been likened to driving a car down the highway without a speedometer. Occasionally, the driver of the car exceeds the speed limit. However, the only way the driver learns of the misdeed is by a police officer stopping the driver to issue a citation. Displaying the dust level to the mining machine operator would be like providing the driver of that car with a speedometer.

What type of information can best be used by the miner? Would instantaneous readout of dust levels help the miner to promptly correct dusty conditions by improving water spray systems, ventilation or mining habits? Or would random short-term high-level excursions of dust levels cause too much confusion? In that case, would a longer integrated value, a running average or a shift average be more appropriate? A record of about a month of shift averages will be stored in the memory incorporated in the MMRDM. Could that information be useful to mine health officials for performing trend analysis to identify mining practices that, on the average, produce less airborne dust? We hope the experience gained during field tests of the prototype MMRDM will answer these questions.

PERSONAL DUST EXPOSURE MONITORS (PDEM)

The need for a PDEM was conceived almost concurrently by researchers at the National Institute for Occupational Safety and Health (NIOSH) and at BOM. For that reason, PDEM development contracts were initiated under an interagency agreement between NIOSH and BOM. Like the MMRDM, the PDEM was originally intended to make instantaneous measurements of airborne respirable coal mine dust and to have the ability to display instantaneous and shift average concentrations. The important differences are that the device must be small enough to be worn by the worker and must have its own power source.

Two types of PDEM are being developed. The first utilizes light scattering as does the MMRDM, but does not make use of a mechanical size separator for respirable dust. The light-scattering configuration is being designed to emphasize the respirable fraction of the aerosol that enters the sensing chamber by convection. The details of this device are covered by Lilienfeld [1983]. If successful, the light-scattering PDEM will provide a rapid-response PDEM to aid workers in controlling their exposure to respirable dust. Equally as important, however, this work will serve to:

1. study the feasibility of measuring respirable coal mine dust using this passive, open-cell, light-scattering approach;
2. provide a relatively inexpensive device that can be used for purposes other than personal exposure monitoring (e.g., a hand-held survey instrument for locating dust sources); and
3. provide a basic building block for other less size- and power-restrictive monitoring applications, since the sensor will necessarily be miniaturized.

The other PDEM will utilize a tapered-element oscillating microbalance (TEOM) technique. The active element of a TEOM consists of an elastic tube constructed of a material with a high mechanical quality factor (Q) and having a special taper. A material with a high Q does not dissipate energy readily, and thus will not damp out quickly when oscillating at a resonant frequency. In the case of the PDEM, the tube is firmly mounted at the wide end while the other end supports an exchangeable filter cartridge. Dust-laden air is drawn through the filter and the resulting filtered air is pumped down the hollow tube. The tapered tube with the filter at the free (narrow) end comprises an oscillating system whose natural frequency will change in relation to the mass deposited on the filter. The tapered element is kept in oscillation by an electrical feedback system. The oscillation of the elastic element is converted into an electrical signal by a light emitting diode (LED)-phototransistor combination. The output of the phototransistor is modulated by the light-blocking effect of the vibrating element.

The TEOM technology originally was intended to be used in a PDEM having the same instantaneous onsite measurement capabilities as the light scattering PDEM. The response time and sensitivity of the technique are adequate. Since the TEOM indiscriminantly measures mass, however, moisture presented a challenging, but not insurmountable, engineering problem. Funding limitations have prevented us from pursuing the con-tractor's suggested design to overcome those problems. Instead, we have opted to obtain six "end-of-shift" prototypes with the monies available.

The end-of-shift concept consists of six TEOM canisters and an electronic readout system. Each TEOM canister will consist of the specially tapered tube with a filter mounted on top, the appropriate electronic hardware to condition the dust sample and tapered element before reading the frequency, and the oscillatory feedback circuitry. This canister will be designed to replace the filter in a typical CMDPSU.

In use, the canister will be placed in a readout system that will:

1. recognize the canister;
2. control the humidity and temperature around the device until the device is in equilibrium with its surroundings;
3. measure and record the resonant frequency of the tapered element system;
4. record the resonant frequency as a type of preweight for the next sampling session; and
5. alert the user if the filter should be changed.

The canister would then be used only as a filter in a CMDPSU during the working shift, i.e., the sampling pump would draw dusty air through the cyclone size selector and deposit respirable dust on the nonoscillating filter. After sampling is finished, the canister would be reinserted into the interface system to determine the change in frequency and thus the weight gain. The filter could be used over and over, each final weight (resonant frequency) serving as the preweight for the next session until the collected mass is too much for the measurement system. The filter would then be replaced.

The development of this PDEM, like the light scattering unit, may provide a very useful PDEM for respirable coal mine dust, and may also serve as a building block for other types of monitors. More importantly, however, we will extensively lab-test the prototypes to determine the capabilities and limitations of a technique for direct measurement of the mass of airborne respirable dust. Light scattering, of course, suffers from the fact that the intensity of the scattered light is not directly proportional to the mass of the particles. In monitoring applications other than PDEM, electrical power could be relatively plentiful. Such power could be used to more easily control moisture problems, and thus provide a means for nearly instantaneous mass measurements.

LIGHT REFLECTANCE

BOM recently concluded a feasibility study of an on-filter light reflectance technique in which the contractor had proposed that the amount of light reflected from a coal-dust-laden filter could be related to the mass of the dust on that filter. The laboratory prototype eventually delivered to BOM made use of a reflective object sensor (ROS) as the light source and detector. A ROS is essentially a LED and phototransistor detector mounted on the same head. The ROS used in the prototype instrument was a commercially available near-infrared device. As a result, the response of the measurement system was strongly dependent on the type of coal dust (anthracite, bituminous, subbituminous, lignite) being viewed. Differences of up to 12% were evident between two different samples of bituminous coal dust. The contractor presented data, obtained using separate yellow LED and detectors, that suggested that a ROS operating in the yellow range would greatly reduce the system's sensitivity to coal dust type. Unfortunately, a yellow ROS in the proper configuration was not commercially available. We may attempt to have such a ROS specially made and test the prototype instrument in-house to determine if the instrument is more acceptable.

Although the prototype light reflectance instrument cannot at present perform acceptably well for most applications, we are now aware of the capabilities and limitations (response time, precision, interferences) of the measurement technique. The technique promises to be simple, durable and probably inexpensive to apply should a measurement need be uncovered that requires a technique with the qualities defined in this study.

EVALUATION OF INSTRUMENTS

A very important activity within the program is the evaluation of dust monitoring instrumentation. Evaluation work serves three purposes:

1. It measures the performance and success of our development contracts.
2. It determines whether or not commercial dust monitors meet manufacturer specifications.
3. It helps determine if a particular monitor is suitable for the mining industry.

Some evaluation work is done in-house. This year, the RAM-1 portable light scattering respirable dust monitor was subjected to extensive aerosol testing in our lab. Two reports will soon be released: The first discusses instrument response to various dusts (two types of coal dust, two types of limestone dust, and Arizona road dust), and the second discusses problems encountered when the instrument is exposed to high concentrations of small water droplets. Such evaluation of the RAM-1 was warranted, because the optical system of the instrument (or a slight modification) is being used in the MMRDM, the light scattering PDEM and other monitoring systems.

Although some evaluation work is conducted in-house, limited facilities and workforce have required that much of the work be contracted. A contract has been awarded to the University of Minnesota to evaluate various dust monitors.

QUARTZ PROJECTS

Another important area of work is the rapid, convenient and accurate measurement of silica (quartz) in dust samples. Two projects in this area of work will be concluded sometime in fiscal year 1982. The first involves the fabrication of a rapid X-ray device to measure quartz. The second project involves developing a method for measuring the quartz content of dust samples that involves the use of an on-filter infrared technique.

MONITORING STRATEGIES

Most of the BOM dust-monitoring program discussed so far consists of projects already underway or completed. As mentioned earlier, these projects were initiated:

1. to satisfy an MSHA request;
2. to study the feasibility of applying a measurement technology to the mining industry; or
3. to develop hardware.

Although these criteria for research and development have served sufficiently well in the past, they can be greatly improved. For example, when studying a novel measurement technique or evaluating an instrument for measuring dust, the researcher can certainly determine such variables as response time, accuracy, precision, sensitivity to interfering agents, noise and durability. The problem arises when deciding, on the basis of the test results, whether or not the device can be applied to the mining industry.

The key word is "application." The need or application, once identified, makes certain requirements on the measurement system. What type of information is needed? Instantaneous readout? Shift averages? Measurements averaged over several days, weeks or months? What limits are required for accuracy? Precision? The important and often difficult step is to properly identify the application. One way to proceed is to make a calculated guess based on past experience as to how a measurement device can best be applied. Occasionally, however, this procedure can result in "gadget" development—instrumentation with little or no real worth or utility in the industry.

We recently initiated a high-priority project to deal with the problem of monitoring applications. Beginning in October 1981, we took the first steps in developing a detailed monitoring strategy for control purposes. We began by examining the mining process to determine those elements that most strongly affect the level of airborne respirable coal mine dust in the mine air. Possible examples might be ventilation, spray water pressure, depth of cut and so on. The most influential elements must be monitored and controlled. The nature of the element (e.g., continuous, intermittent, random) will define the quality and nature of data that is needed. Once these needs are ascertained, one can refer to existing technology and available instrumentation to see if they can satisfy the monitoring needs defined by the monitoring strategy. When the strategy identifies a monitoring need for which there is no existing hardware, then hardware development can be easily justified and the specifications for the device will be clearly definable.

Development of the strategy for control monitoring purposes will not be trivial and is expected to extend over several years. The description of the work has, of course, been greatly simplified; however, the importance of such a strategy must not be underemphasized.

REFERENCES

Lilienfeld, P. (1983) "Current Mine Dust Monitoring Developments," in *Aerosols in the Mining and Industrial Work Environment, Vol. 3, Instrumentation*, V. A. Marple and B. Y. H. Liu, Eds. (Ann Arbor, MI: Ann Arbor Science Publishers).

CHAPTER 10

NATIONAL INSTITUTE FOR OCCUPATIONAL SAFETY AND HEALTH AEROSOL RESEARCH RELATING TO MEASUREMENT AND REGULATION

P. A. Baron

Division of Physical Sciences and Engineering
National Institute for Occupational Safety and Health
Cincinnati, Ohio

ABSTRACT

Aerosol research currently carried out by various divisions within the National Institute for Occupational Safety and Health (NIOSH) is described. The areas of research include toxicology, epidemiology and associated field studies, sampling and analytical methods, and instrumentation. Some areas for future research also are presented.

INTRODUCTION

NIOSH, in pursuing its mission to protect and improve the health of workers in mining and industrial workplaces, conducts research in several fields that require the use of aerosol measurement and generation. This research is carried out by different divisions within NIOSH and forms the basis for recommendations to the Occupational Safety and Health Administration (OSHA) and the Mine Safety and Health Administration (MSHA) on criteria for health standards. While much work is carried out in-house, a considerable portion is carried out under contracts, grants and interagency agreements.

The aerosol-related research within various research divisions of NIOSH will be presented, not in an attempt to be comprehensive, but to give an overview of the types of work being carried out. For further detail, the list of references or the "NIOSH Publications Catalog" [NIOSH 1980] can be consulted. Various aspects of possible future aerosol research will also be discussed.

CURRENT STUDIES

Division of Standards Development and Technology Transfer

This division has the responsibility to compile information available from the literature and NIOSH research and to formulate the official criteria for recommended standards that are forwarded to the regulatory agencies OSHA and MSHA. These documents provide the best available recommendations on sampling and analytical techniques, and point out areas needing further research. Past aerosol-oriented criteria document topics have included silica, asbestos, carbon black, inorganic nickel and lead [NIOSH 1975,1977a,b, 1978a,b]. Currently, a document on asbestos in mines is in preparation for MSHA.

Division of Biomedical and Behavioral Sciences

This division carries out toxicological research on various air contaminants. Table I contains a list of materials with which animal inhalation exposure testing has recently been carried out or is currently underway [Groth et al. 1981, in press; Mackay et al. 1980; Moorman et al. submitted, in preparation; Palmer et al. 1980, in press, submitted; Stettler et al. in preparation]. It is interesting to note that >60% of the inhalation exposure experiments were carried out using the Wright dust feeder. Although this device has many advantages for long-term exposure experiments, it does not produce an ideal, well dispersed aerosol. Other generators used include nebulizers, chain-fed aspirators and fluidized beds. Most of the studies in the table use electron microscopy for particle sizing and chemical analysis. A description of these

Table I. Workplace Pollutants Inhalation Exposure Experiments

Polyvinyl Chloride	Asbestos (short fiber chrysotile)
Amorphous silica	$AlCl_3/AlF_3/KF$
Bituminous coal	Polyurethane foam
Fibrous glass	Bisphenol A
Coal dust/diesel fume	Platinum salts (hexachloroplatinate)

techniques is given by Stettler et al. [1983]. The electron microscope analysis was augmented by bulk chemical analysis techniques in some cases.

Division of Respiratory Disease Studies

This division performs clinical and epidemiological research related to occupational respiratory disease. Several ongoing studies within this division include a volcanic ash exposure study as a followup to the Mt. St. Helens eruption, a joint NIOSH–U.S. Department of Agriculture long-term human exposure study on washed cotton dust and an evaluation of immune response to organic dusts in the logging industry. Other studies include industrial hygiene/epidemiological investigations of exposure to wood, grain, diesel/ coal, crushed stone and nonasbestos fiber dusts.

Division of Surveillance, Hazard Evaluation and Field Studies

The primary research functions of this division lie in the areas of epidemiology and field studies. Most of the studies carried out by this division attempt to characterize all the important pollutants in the workplace. Aerosol measurements may represent only part of the monitoring carried out. There are a variety of ongoing studies, including a series of mortality and industrial hygiene studies on aluminum-reduction workers exposed to fluorides, alumina and coal tar pitch volatiles; leather workers and others exposed to azo dyes; workers in a variety of industries exposed to sulfuric acid; and automotive wood die and model makers exposed to wood dusts and glues.

Division of Physical Sciences and Engineering

This division is primarily responsible for providing measurement and analytical techniques for assessing pollutants in the workplace. There are several branches, including the Methods Research Branch, which is responsible for sampling and analytical methods. Recent projects in this branch have included analysis of silica by X-ray diffraction (XRD), inorganic acids by ion chromatography, welding fumes by X-ray fluorescence (XRF) and asbestos by XRD and differential thermal analysis. Methods such as these are included in the yearly updates to the NIOSH "Manual of Analytical Methods" [Taylor 1977a,b,c,1978,1979,1980]. There is a collaborative test being carried out to evaluate XRD and infrared methods for silica.

The Monitoring and Control Research Branch performs research on field instrumentation and evaluates and develops engineering control techniques. Current projects in engineering controls include the development of a push-pull ventilation system. A survey of control monitoring instrumentation has recently been carried out, and a report soon will be available. This report provides an extensive compilation of dust as well as gas and vapor control monitoring instrumentation and describes applications of current technology in control monitors for the workplace atmosphere [Bochinski 1981].

The Monitoring Research Section has been the group traditionally most involved in aerosol monitoring and instrumentation. Areas of special interest have been respirable dust sampling, direct reading instrumentation and asbestos measurement techniques. Some recently completed and current projects include:

1. Development of an aerodynamic particle sizer (APS): this instrument measures aerodynamic size distributions in the range of 0.8–15 μ in near-real-time. APS prototype testing was carried out at NIOSH and will be reported by Baron [1983] and Remiarz et al. [1983]. The APS is now being marketed commercially by TSI, Inc.

2. Development of an XRF analyzer: this device is a portable, nondestructive field analyzer of filter samples for >20 elements. Two prototypes have been received by NIOSH. Testing will take place during the next year. (Development was a joint effort with the Department of Energy.)

3. Development of a personal dust exposure monitor: two approaches to a personal dust monitor are being investigated under a joint effort with the Bureau of Mines. These two approaches, one using a light scattering detector and the other a tapered-element oscillating microbalance, are being developed under separate contracts. The light scattering approach is described by Lilienfeld [1983].

4. Certification procedures for the coal mine dust personal sampler: an investigation of the precision and accuracy of measurements made with the 10-mm nylon cyclone with associated pump and filter holder has produced new recommendations for the certification of coal mine dust sampler units. Certification of coal mine dust personal sampler units is carried out by NIOSH in the Testing and Certification Branch in Morgantown, West Virginia. This information is being prepared as part of a report for MSHA. A description of a technique for modeling cyclone penetration curves is presented by Bartley et al. [1983].

5. Evaluation of sampling strategies for coal mine dust: mathematical evaluation of dust sampling strategies is presented by Bowman et al. [1983].

6. Inlet effects in total dust samplers: several studies have been carried out. A contract at IIT Research Institute has produced a report and a computer program on theoretical inlet efficiencies [Rajendran 1979]. An interagency agreement with Los Alamos National Laboratory has produced a report on experimental inlet efficiencies [Fairchild et al. 1980]. A grant to the University of Cincinnati has produced results reported by Willeke and Tufto [1983].

It should be noted that there is extensive interaction between the various divisions within NIOSH. For instance, analytical methods are developed and applied for field sampling projects. Instrumentation is developed and modified for field use and for toxicology experiments. Field studies indicate what research is needed in control technology. The results of toxicology experiments indicate the levels and types of pollutants requiring sampling and analytical methods. Finally, all of this research provides input into new criteria documents and the development of criteria documents indicates research needs or gaps in the various areas.

There are a number of areas where further development of aerosol generation and measurement are needed. The following comments are some ideas and directions for future research.

In the area of inhalation toxicology, many aerosols are generated from preground materials, using generators such as the Wright dust feeder. These aerosols can have properties very different from those generated in the workplace. For example, the size distribution may have shifted, the particles may be more agglomerated, the charge distribution may be different, and the surface properties may have changed. The interpretation of the toxicity of aerosols is difficult enough without introducing confounding variables. New techniques are needed to simulate the generation processes that actually produce the aerosols in the workplace. The use of a diesel engine to produce diesel fumes for inhalation toxicity testing is a good example of this approach.

Improved characterization of aerosol exposure experiments through continuous monitoring is now possible. Commercially available instrumentation can be used to monitor mass levels and size distributions continuously, rather than depending on periodic measurement with filter samples and cascade impactors. It may be possible to adapt control monitoring instrumentation to exposure chambers to adjust and control the aerosol concentrations.

Field instrumentation, used to augment traditional sampling methods, has advanced tremendously in the last ten years. However, there still seems to be reluctance among industrial hygienists to use this equipment. The reasons are varied, but the major factors seem to be cost, accuracy, portability and ruggedness.

Field analytical instrumentation is an area that is just starting to develop. In gas and vapor monitoring, for example, there are commercially available field units using gas chromatography, colorimetry and electrochemistry. The portable XRF instrument mentioned above will allow the industrial hygienist to quickly assess the levels of specific pollutants in aerosols.

In the area of sampling, two subjects of current concern include sampler inlet efficiencies and electrostatic sampler–aerosol interaction. As mentioned above, several studies have been done on inlet efficiencies under different air velocities and directions. Considerably more work remains to be done, especially in defining which particles larger than 10 μ are, or should be, sampled when measuring total dust.

REFERENCES

Baron, P. A. (1983) "Sampler Evaluation with an Aerodynamic Particle Sizer," in *Aerosols in the Mining and Industrial Work Environment, Vol. 3, Instrumentation*, V. A. Marple and B. Y. H. Liu, Eds. (Ann Arbor, MI: Ann Arbor Science Publishers).

Bartley, D. L., J. D. Bowman, G. M. Breuer and L. J. Doemeny (1983) "Accuracy of the 10-Millimeter Cyclone for Sampling Respirable Coal Mine Dust," in *Aerosols in the Mining and Industrial Work Environment, Vol. 3, Instrumentation*, V. A. Marple and B. Y. H. Liu, Eds. (Ann Arbor, MI: Ann Arbor Science Publishers).

Bochinski, J. H. (1981) "Assessment of Engineering Control Monitoring Equipment," Final Report, NIOSH Contract 210-79-0011.

Bowman, J. D., K. A. Busch and S. A. Shulman (1983) "Statistical Evaluation of Sampling Strategies for Coal Mine Dust," Chapter 23, this volume.

Fairchild, C. I., M. I. Tillery, J. P. Smith and F. O. Valdez (1980) "Collection Efficiency of Field Cassettes," Los Alamos Technical Report LA-8640-MS.

Groth, D. H., W. J. Mooreman, D. W. Lynch, L. E. Stettler, W. D. Wagner and R. W. Hornung (1981) "Chronic Effects of Inhaled Amorphous Silicas in Animals," in *Health Effects of Synthetic Silica*, D. D. Dunnom, ASTM STP 732 (Philadelphia: American Society for Testing and Materials).

Groth, D. H., D. W. Lynch, W. J. Moorman, C. Kommeneni, L. E. Stettler, T. R. Lewis and W. D. Wagner (in press) "Pneumoconiosis in Animals Exposed to Polyvinyl Chloride Dust," *J. Environ. Persp.*

Lilienfeld, P. (1983) "Current Mine Dust Monitoring Developments," in *Aerosols in the Mining and Industrial Work Environment, Vol. 3, Instrumentation*, V. A. Marple and B. Y. H. Liu, Eds. (Ann Arbor, MI: Ann Arbor Science Publishers).

Mackay, G. R., L. E. Stettler, C. Kommenini and H. M. Donaldson (1980) "Fibrogenic Potential of Slags Used as Substitutes for Sand in Abrasive Blasting Operations," *Am. Ind. Hygiene Assoc. J.*

Moorman, W. J., J. B. Lal, R. E. Biagini and W. D. Wagner (submitted) "Pulmonary Effects of Chronic Exposure to Polyurethane Foam (PUF)," *Am. Ind. Hygiene Assoc. J.*

Moorman, W. J., T. R. Lewis, W. D. Wagner, C. Kommenini and D. H. Groth (in preparation) "Pulmonary Responses to Long-Term, Low-Level Bituminous Coal Exposure."

NIOSH (1975) "Criteria for a Recommended Standard: Occupational Exposure to Crystalline Silica," DHEW Report (NIOSH) 77-120.

NIOSH (1977a) "Criteria for a Recommended Standard: Occupational Exposure to Asbestos (Revised)," DHEW Report (NIOSH) 77-169.

NIOSH (1977b) "Criteria for a Recommended Standard: Occupational Exposure to Inorganic Nickel," DHEW Report (NIOSH) 77-164.

NIOSH (1978a) "Criteria for a Recommended Standard: Occupational Exposure to Carbon Black," DHEW Report (NIOSH) 78-204.

NIOSH (1978b) "Criteria for a Recommended Standard: Occupational Exposure to Inorganic Lead (Revised)," DHEW Report (NIOSH) 78-158.

NIOSH (1980) "NIOSH Publications Catalog," DHHS Report (NIOSH) 80-126.

Palmer, W. G., R. C. Scholz and W. J. Moorman (1980) "Carcinogenic Potential of Condensed Pyrolysis Effluents from Iron Casting Operations—Preliminary Report," *Am. Found. Soc. Trans.* 80-65:745.

Palmer, W. G., R. H. James, W. J. Moorman, R. C. Scholz and L. Stettler (in press) "Analysis of Effluents Collected from Four Types of Iron Foundry Casting Molds for Use in Carcinogenesis Bioassay," *Am. Found. Soc. Trans.*

Palmer, W. G., R. C. Scholz and W. J. Moorman (submitted) "An Innovative Approach to Sampling Complex Industrial Emissions for Use in Animal Toxicity," *Am. Ind. Hygiene Assoc. J.*

Rajendran, N. (1979) "Theoretical Investigation of Inlet Characteristics for Personal Aerosol Samplers," Final Report, NIOSH Contract 210-78-0092.

Remiarz, R. J., J. K. Agarwal, F. R. Quant and G. J. Sem (1983) "A Real-Time Aerodynamic Particle Size Analyzer," in *Aerosols in the Mining and Industrial Work Environment, Vol. 3, Instrumentation*, V. A. Marple and B. Y. H. Liu, Eds. (Ann Arbor, MI: Ann Arbor Science Publishers).

Stettler, L. E., S. F. Platek and D. H. Groth (1983) "Particle Analysis by Scanning Electron Microscopy—Energy-Dispersive X-Ray Analysis—Image Analysis," in *Aerosols in the Mining and Industrial Work Environment, Vol. 3, Instrumentation*, V. A. Marple and B. Y. H. Liu, Eds. (Ann Arbor, MI: Ann Arbor Science Publishers).

Stettler, L. E., H. M. Donaldson and G. C. Grant (in preparation) "Chemical Composition of Coal and Other Mineral Slags."

Taylor, D. (1977a) "NIOSH Manual of Analytical Methods, Vol. 1," DHEW Report (NIOSH) 77-157a.

Taylor, D. (1977b) "NIOSH Manual of Analytical Methods, Vol. 2," DHEW Report (NIOSH) 77-157b.

Taylor, D. (1977c) "NIOSH Manual of Analytical Methods, Vol. 3," DHEW Report (NIOSH) 77-157c.

Taylor, D. (1978) "NIOSH Manual of Analytical Methods, Vol. 4," DHEW Report (NIOSH) 78-175.

Taylor, D. (1979) "NIOSH Manual of Analytical Methods, Vol. 5," DHEW Report (NIOSH) 79-141.

Taylor, D. (1980) "NIOSH Manual of Analytical Methods, Vol. 6," DHEW Report (NIOSH) 80-125.

Willeke, K., and P. Å. Tufto (1983) "Sampling Efficiency Determination of Aerosol Sampling Inlets," Chapter 25, this volume.

PART 2

PARTICLES AND THE RESPIRATORY SYSTEM

CHAPTER 11

ROLE OF PARTICLE DEPOSITION IN OCCUPATIONAL LUNG DISEASE

Morton Lippmann, Joshua Gurman
and Richard B. Schlesinger

Institute of Environmental Medicine
New York University Medical Center
New York, New York

ABSTRACT

Inhaled particles that deposit at various specific sites within the respiratory tract can initiate lesions at those sites, which can contribute to acute or chronic disease. The pattern of deposition is also important for those particles that are cleared before causing damage at the deposition site. The deposition sites determine the clearance pathways, and particles can accumulate at and cause injury to sites along these pathways. Quantitative aspects of the deposition of inhaled particles within the major functional regions of the human respiratory tract are reviewed and discussed in relation to recommendations for size-selective aerosol sampling. Deposition within the tracheobronchial region is discussed further, with specific emphasis on the variations in deposition within the larger bronchi.

INTRODUCTION

Airborne particles in the breathing zone can be inhaled and deposited at various sites along the conductive airways and gas-exchange regions of the respiratory tract. The particles that deposit at the various sites can initiate or exacerbate a biological response at the site of deposition, which, in turn, can lead to clinically defined lung disease. The particles also can be translocated from their sites of deposition to other sites within or beyond the respiratory tract by clearance processes, and can therefore reach other tissues, where disease processes can take place.

Silicosis and coal worker's pneumoconiosis are associated with particles that deposit in the nonciliated alveolar portions of the lungs; bronchial cancers, with particles that deposit on the larger bronchial airways; and chronic bronchitis with particles that deposit on intermediate and smaller ciliated lung airways. Although the incidence and progression of these diseases are affected by other factors, such as the functioning of host defenses, proper evaluations of disease potential of current exposures and controls depend on quantitative knowledge of the amount and pattern of particle deposition within the component regions of the human respiratory tract.

The inhaled fraction of the airborne particles depends on the aerodynamic size distribution of the aerosol and the inlet characteristics of the respiratory tract; this aspect is discussed in detail by Armbruster et al. [1982] and Ogden [1982]. The fraction of the inhaled particles that deposit in the major functional regions of the respiratory tract (head, tracheobronchial, alveolar) depends on:

- aerodynamic diameter (for particles larger than 0.5 μm),
- linear dimensions (for submicrometer aerosols and fibers),
- hygroscopic growth within the airways,
- particle electrical mobility,
- respiratory flow patterns, rates, volumes and direction, and
- airway and airspace sizes.

In addition to further discussion in this chapter these topics are addressed by Schlesinger et al. [1982]; Martonen and Lowe [1982]; Ogden [1982]; Swift [1982]; Yu and Hu [1982].

The effects of deposited particles on respiratory tract airways may depend on the pattern of deposition at specific sites and the amount deposited in each region as a whole. Within larger airways, it has been observed that the sites where the particle deposition is greatest (e.g., the larynx and the bifurcations in the central bronchi) are also the sites where primary cancers are most commonly found [Schlesinger and Lippmann 1972,1978].

Since 1970 the Institute of Environmental Medicine has used hollow airway casts for quantitative determination of deposition within specific airways of the tracheobronchial tree, providing the first detailed quantitative descriptions of the effects, in a realistic airway geometry, of:

1. jet-flow through the larynx on deposition pattern within the trachea on the tracheal bifurcation and deposition efficiency in subsequent airway generations [Chan et al. 1980; Schlesinger and Lippman 1976; Schlesinger et al 1977];
2. particle size, particle electrical charge, inspiratory flowrate and flow pattern on deposition efficiencies within the trachea and bronchi during inspiratory flow [Chan et al. 1978; Schlesinger et al. 1977; Schlesinger et al. in press]; and
3. particle size and flowrate on deposition patterns and efficiencies within the trachea and larger bronchi during expiratory flow.

These kinds of data cannot be obtained from in vivo tests, where the airways are inaccessible for detailed measurements and where clearance processes would move the particles before they could be measured. Nor can they be obtained from tests in idealized tubular airway models, where the walls are smooth, and where the cross-sectional changes at and near the airway bifurcations cannot realistically be simulated [Chan et al. 1980; Schlesinger 1980]. The advantages of hollow airway casts for deposition studies have been recognized by others. For example, we supplied casts to Martonen and Lowe [1982] to study the deposition of aerosols in the bronchial airways. Timbrell [1972] used hollow pig lung casts to study fiber penetration through the bronchial airways.

Other particle concentrations within the lungs that are associated with occupational diseases result from particle translocation and accumulation, rather than from direct deposition. These include dust foci around respiratory bronchioles, and concentrations within lymph nodes. However, these non-deposition-related factors are beyond the scope of this chapter. They were described in detail in a review paper [Lippmann et al. 1980].

REGIONAL DEPOSITION

Available data on the regional deposition of inhaled particles in the human respiratory tract have been summarized by the U.S. Environmental Protection Agency (EPA) in their draft criteria document for particulate matter and sulfur oxides. The data considered reliable for deposition in the head (extrathoracic), tracheobronchial (TB) tree and nonciliated pulmonary (alveolar) regions of healthy humans are summarized in Figures 1 to 4. There is a great amount of intersubject variability in deposition in all regions, due both to their inherent variability in airway and airspace dimensions, and the variability in breathing rates and patterns. Even so, there is a clear particle size dependence. Deposition in the head is primarily by impaction, and a comparison of Figures 1 and 2 shows that the nasal passages are much more efficient particle collectors than the oral passages. In the TB airways, impaction is the dominant removal mechanism for particles larger than about 2 μm under most conditions, while sedimentation is the major collection mechanism for particles between ~0.5 and 2 μm. As an impactor, the TB region is much more efficient than the oral airways, but somewhat less efficient than the nasal airways. Thus, particle deposition within the lungs for 1- to 10-μm particles is very much dependent on whether the individual breathes through the nose or mouth. Deposition in the pulmonary region is primarily by sedimentation for particles larger than 0.5 μm, and by diffusion for smaller particles. Particles of ~0.5 μm, having a minimal intrinsic mobility, have a minimum in deposition probability. For particles larger than ~3 μm there is less pulmonary deposition with increasing size, because these larger particles have a diminishing penetration through the conductive airways.

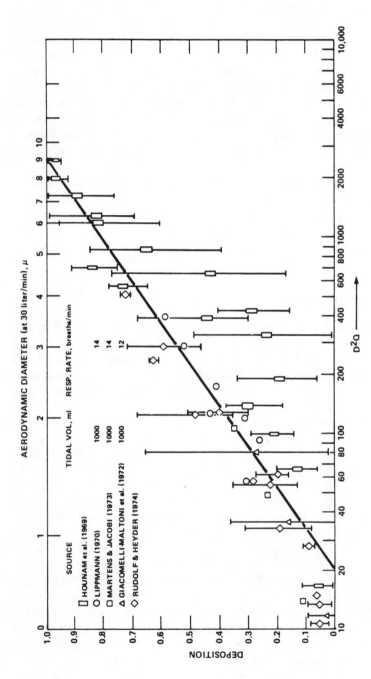

Figure 1. Deposition of monodisperse aerosols in the extrathoracic region for nasal breathing in humans as a function of D^2Q, where Q is the average inspiratory flowrate in liter/min. The solid line is the International Commission on Radiological Protection deposition model based on data of Pattle (1961). Other data show the median and range of the observations as cited by the various investigators. (From EPA Draft Criteria Document for Particulate Matter and Sulfur Oxides.

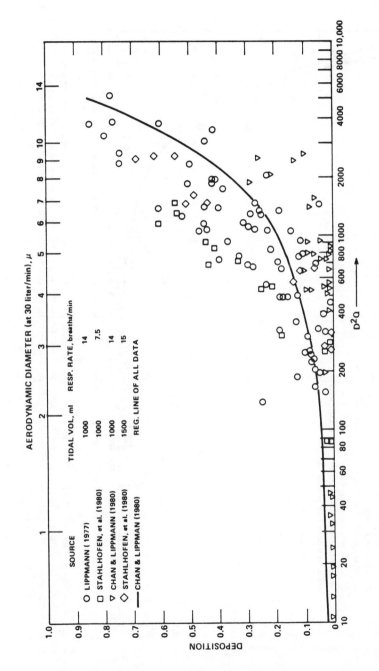

Figure 2. Deposition of monodisperse aerosols in extrathoracic region for mouth breathing in humans as a function of D^2Q where Q is the average inspiratory flowrate in liter/min. The data are the individual observations as cited by the various investigators. The solid line is the overall regression derived by Chan and Lippmann [1980]. (From EPA Draft Criteria Document.)

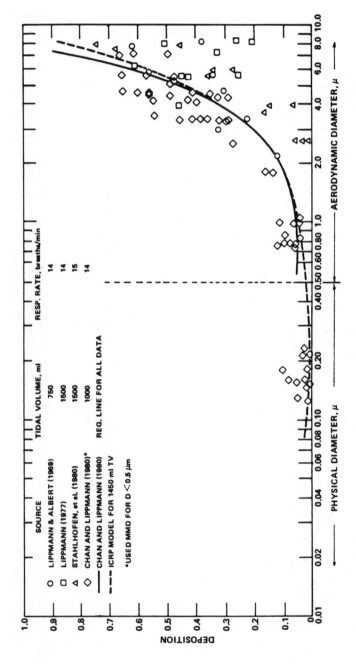

Figure 3.   Deposition of monodisperse aerosols in the TB region for mouth breathing in humans in percent of the aerosols entering the trachea as a function of aerodynamic diameter (except below 0.5 μm where deposition is plotted vs physical diameter as cited by different investigators). Dashed line is ICRP model for 1450-ml tidal volume. The solid line is the overall regression derived by Chan and Lippmann [1980]. (From EPA Draft Criteria Document.)

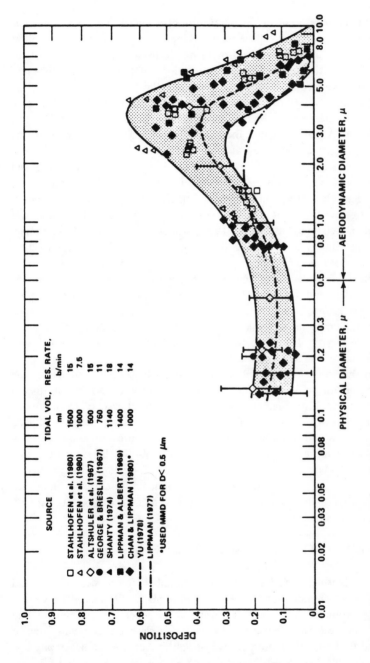

Figure 4. Deposition of monodisperse aerosols in the pulmonary region for mouth breathing in humans as a function of aerodynamic diameter (except below 0.5 μm, where deposition is plotted vs physical diameter). The eye-fit band envelops deposition data cited by different investigators. The dashed line is the theoretical deposition model of Yu [1978] and the broken line is an estimate of pulmonary deposition for nose breathing derived by Lippmann [1977]. (From EPA Draft Criteria Document.)

DEPOSITION WITHIN SPECIFIC TB AIRWAYS

Experimental measurements of deposition efficiencies and the surface density of deposited particles can be made, as indicated previously, using hollow airway casts of human lungs, and/or using idealized physical models of the airways. The relative advantages and limitations of both casts and physical models were discussed in detail by Schlesinger [1980].

Studies using idealized models have shown the influence of various physical factors on TB deposition. Martin and Jacobi [1972] exposed two Weibel [1963] Model A plastic tubular models of the trachea through the segmental bronchi, which were connected at their distal ends with tubing, to 0.2- to 0.4-μm particles tagged with ^{212}Pb. All of the airways were in one plane, and no larynx was used. Deposition efficiencies in inhalation and exhalation were similar at high flowrates, and turbulent diffusion had a major impact on efficiency. At low flowrates, the inspiratory/expiratory deposition ratio was much higher. Bell and Friedlander [1973] studied deposition in a two-dimensional wedge model and a three-dimensional tube model of the tracheal bifurcation, both with no upstream larynx, and found up to a fourfold enhancement in deposition for cyclic as compared to constant flow. On the other hand, Lee and Wang [1977] found little difference between inspiratory and expiratory flow deposition efficiencies in smooth-walled glass tube models of the larger bronchial airways having a venturi section at the inlet to simulate the larynx. They also found that deposition efficiency varied with the orientation of the branching airways.

Work in this laboratory on hollow cast systems began with the production of casts extending to 5-mm-diameter airways. Our subsequent studies used a cast extending to 2-mm-diameter airways, and utilized casts of the larynx to produce more realistic entry conditions. Our most recent studies have utilized replicate casts, which could be cut up into specific bifurcation and length segments, allowing more precise measurement of distribution by segment as well as microscopic analysis of the surface density of deposition in the vicinity of the bifurcations. In these studies, we also examined differences in deposition between constant and cyclic inspiratory flows, and deposition during constant expiratory flow.

Influence of Cyclic Flow on Deposition Efficiency and Pattern

Gurman performed a series of hollow cast inhalation studies with 3- and 8-μm mass median aerodynamic diameter (MMAD) ^{99m}Tc-tagged Fe_2O_3 microspheres produced by a spinning-disc aerosol generator [Lippmann and Albert 1967]. Tests were performed for each particle size at constant flows of 15, 30 and 60 liter/min, representative of light, moderate and heavy physical activity, and with cyclic inspiratory flows that averaged 15, 30 and 60 liter/

min. All tests were performed using a variable-orifice larynx [Gurman et al. 1980], with the orifice held at a suitable constant size for the constant-flow tests. Each test was performed with a replicate hollow airway cast extending to 3-mm-diam. airways, which was coated on the inside with a thin layer of silicone oil to simulate airway mucus. Each replicate was made from the same solid master TB cast, which was itself made from a normal human lung. Each cast was cut up, after the inhalation, in appropriate airway bifurcation and length segments, permitting accurate quantification of deposition efficiency in each segment by gamma-counting in a scintillation well counter, and of deposition pattern at bifurcations by microscopic counting of particles. The results of the surface density analyses will be discussed in the next section.

Figure 5 shows the summary of deposition efficiency by airway generation for all of the 3-μm tests, and Figure 6 for the 8-μm tests. It can be seen that deposition efficiency at each generation increases with both flowrate and particle size, as would be expected for this system and these particle sizes, where

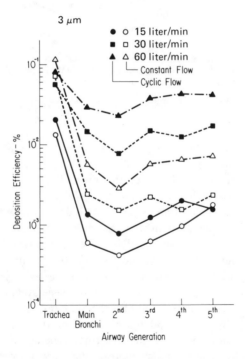

Figure 5.   Deposition efficiencies in each airway generation of replicate hollow airway casts of a human lung for a monodisperse 3-μm MMAD aerosol for constant and cyclic inspiratory flows at three different rates.

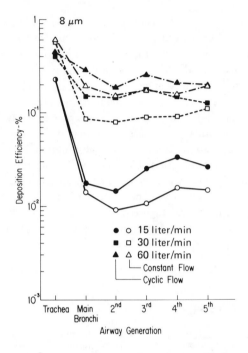

Figure 6. Deposition efficiencies in each airway generation of replicate hollow airway casts of a human lung for a monodisperse 8-μm MMAD aerosol for constant and cyclic inspiratory flows at three different rates.

inertial effects have a major influence. It can also be seen that deposition efficiencies are substantially higher, in all of the bronchial airways, for cyclic flow than for constant flow. However, the reverse is true for deposition in the trachea for the 3-μm particles at 30 and 60 liter/min.

Within the trachea, deposition per unit length decreased with distance from the larynx for all of the constant flow tests at both 3 and 8 μm, with a much more rapid decline for the 3-μm particles. The pattern was similar for cyclic flow at 15 liter/min. However, for the 30- and 60-liter/min tests, there was a strong secondary peak in deposition at 5–6 cm below the larynx. In the bronchial airways, the average surface deposition per generation exhibited similar patterns for all of the 3-μm tests, and for the 8-μm tests at 15 liter/min. However, for the 8-μm particles at 30 and 60 liter/min, there was a tendency for more of the deposition to be in the first three generations in the constant-flow tests, and there was a much more pronounced proximal shift in the cyclic-flow tests.

In previous deposition studies conducted in this laboratory, the results of various tests using particles >2 μm MMAD, where particle momentum affects deposition, have been normalized by relating deposition efficiency to Stokes number [Chan and Lippman 1980]. Stokes number is a dimensionless term representing the ratio of the stopping distance of the particle, due to the viscous drag of the fluid, to the characteristic dimension of the system. In the current study, when the deposition efficiencies in each airway generation for individual test runs were correlated with Stokes number, all of the values of the correlation coefficient (r), for both constant and cyclic flows, were 0.98 or better. However, when the data for each run from all generations were combined, the overall r values were only 0.90 for constant flow and 0.84 for cyclic flow. Some of the differences among the airway generations that contributed to the reduced correlation coefficients could be accounted for by inserting a branching angle term into the correlation, bringing the r values up to 0.92 for constant flow and 0.88 for cyclic flow.

Most of the remaining variability between airway generations was found to be associated with the downstream projected area of the impaction surface below the jet, i.e., the projected area of the bronchus. A combined term consisting of the Stokes number times the size of the branch angle times the projected bronchial surface area, when correlated with the deposition efficiency, yielded r values of 0.98 for constant flow and 0.97 for cyclic flow. The results are plotted in Figure 7 for constant flow and in Figure 8 for cyclic flow.

Figure 8 shows that the deposition efficiency is flowrate-dependent in cyclic flow. As compared to the constant-flow relationship from Figure 7, there is enhanced deposition in cyclic flow averaging 15 liter/min, and substantially more enhancement at 30 liter/min. There appears to be some further enhancement at 60 liter/min, but this is less firmly established. The enhanced deposition in cyclic flow is most likely related to persistence of turbulence associated with the peak flows reached during the flow cycle. The differences diminish as particle size increases, and for the flowrates above 30 liter/min, because the intrinsic mobility of the particle affects deposition under these conditions more than does its movement in turbulent eddies.

Deposition Pattern at the Tracheal Bifurcation

The surface density of deposition was mapped on the bifurcation between the trachea and main bronchi. This bifurcation section was slit laterally to flatten it, and then further cut into three sections, which were mounted onto microscope slides. Each section was scanned using a binocular microscope at 100X with a grid in the eyepiece to allow delineation of the field into 1-mm^2

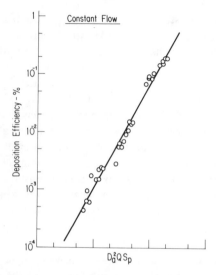

Figure 7.   Deposition efficiencies for constant inspiratory flow in the first five branching levels of hollow airway casts of a human lung as a function of an optimized
normalization factor $D_a^2QS_p$, where D_a is the aerodynamic particle diameter, Q is
the inspiratory flowrate and S_p is the projected area of the bronchus below the
jet. The data points are for five branching levels, two particle sizes (3 and 8 μm)
and three flowrates (15, 30 and 60 liter/min).

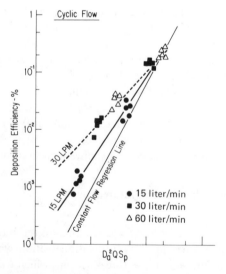

Figure 8.   Deposition efficiencies for cyclic inspiratory flow for the same replicate
hollow airway casts, particle sizes and flowrates as in Figure 7. It can be seen that
the efficiencies are higher in cyclic flow than in constant flow, except for the 8-μm
particles at 60 liter/min.

areas. The number of particles within each area was counted optically in selected regions, namely, the center of the carinal ridge, 1 mm on each side of
the carina and every other 1-mm section over the entire bifurcation area. The
observed densities within these areas were corrected for actual area, since the
sections were not optically flat when viewed in the microscope field.

Figure 9 presents maps of the surface distribution of particles at the first
bifurcation for 8-μm particles under 15- and 60-liter/min cyclic and constant
flows. Constant flow results in denser deposition along the carinal ridge,
especially at the lower flowrate. In addition, greater particle density occurs
towards the front (ventral) wall of the airway under constant flow, while under variable flow, the tendency is for greater surface density towards the rear
(dorsal) wall. Thus, the epithelial cells at the carinal edge may be expected to
receive a particle dose greater by an order of magnitude than that received by
others less than 1 cm away. The degree of enhancement in surface density of
deposition is, however, quite variable, being dependent on the flowrate and
its temporal variation.

Particle Deposition During Exhalation in Casts of the Human TB Tree

Very little information is available regarding the deposition of inhaled particles in the human TB tree during exhalation. A simple dual hollow cast system was developed to initiate studies of TB deposition during exhalation, and
are described by Schlesinger et al. [1982]. The major findings of this study
were:

1. Exhalation deposition is less efficient than inhalation deposition, with the
    ratio of exhalation to inhalation deposition efficiency varying from 0.62 to
    0.82.
2. Deposition in the inhalation cast was enhanced (1) for several centimeters
    distal from the larynx, and (2) at the tracheal bifurcation.
3. Exhalation cast deposition is enhanced (1) for several centimeters proximal
    to the tracheal bifurcation, and (2) in the main bronchi proximal to the main
    bronchial bifurcations. Apparently, particle deposition is enhanced down-
    stream of bifurcations during both inhalation and exhalation.

AEROSOL SAMPLING FOR HAZARD EVALUATIONS

Since the inhalation hazard associated with airborne particles is dependent
on where they deposit within the respiratory tract, it is appropriate to separate the particles into aerodynamic size fractions corresponding to their
expected regional deposition, and to establish exposure concentration limits
for the specific size fractions associated with the health effects of interest. In
1952 the British Medical Research Council (BMRC) established sampling
criteria for coal mine dust based on their recognition that coal worker's
pneumoconiosis was caused by that fraction of the mine dust that penetrated

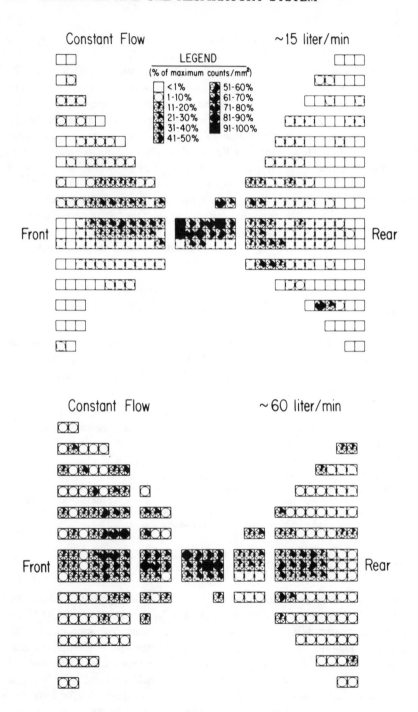

Figure 9. Surface density maps at the bifurcation of the trachea in hollow airway

Cyclic Flow   ~15 liter/min

Front   Rear

Cyclic Flow   ~60 liter/min

Front   Rear

casts for 8-μm MMAD particles for constant and cyclic flows of 15 and 60 liter/min.

through the conductive airways and could deposit in the pulmonary region. They named this fraction "respirable" dust, and defined a sampler acceptance criterion based on the performance of a horizontal elutriator that collected 100% of 7.1-μm and 50% of 5-μm particles. In 1960 the Atomic Energy Commission (AEC) used the same rationale for dusts in the uranium milling industry, but defined a slightly different sampler acceptance criterion, i.e., 50% capture at 3.5 μm, and 100% collection at 10 μm. In 1968 the American Conference of Governmental Industrial Hygienists (ACGIH) adapted the AEC sampling criteria for their threshold limit values (TLV) for coal mine and silica dusts. "Respirable" dusts, as defined by BMRC, AEC and ACGIH, relate only to diseases of the pulmonary region. Particles that deposit in the head and TB regions are thus, by their definitions, nonrespirable.

More comprehensive definitions are needed for particles that deposit in the head and TB regions, and cause diseases such as nasal and bronchial cancers and chronic bronchitis. Miller et al. [1979] of the EPA defined "inhalable" dust as that having $\geqslant$10% penetration into the trachea during mouth breathing. They defined the D_{50} for such particles as 15 μm. A working group appointed by the International Standards Organization [ISO 1981] defined a series of aerosol fractions. The fraction drawn in by the nose or mouth was called "inspirable," that part collected in the head was called "extrathoracic," and that part penetrating through the larynx was called "thoracic" and was further subdivided into "tracheobronchial" and "alveolar." Their recommendations were presented with two options. One, with a D_{50} cut of 15 μm for penetration into the trachea and one with a D_{50} cut of 10 μm. The 15-μm cut would be consistent with Miller et al. [1979] and be conservative for lung disease. The 10-μm cut (Figures 10 and 11) would give a more unbiased cut between head and chest fractions.

In the ballotting earlier this year, 88% of the national delegations voted for the 10-μm cut size. Fortuitiously, at least in terms of uniformity of practice, in July 1981 the EPA Office of Air Quality Planning and Standards recommended to the Administrator that the revised particulate matter standard should include a D_{50} of 10 μm. The fraction below the 10-μm cut, designated total thoracic particulate (TTP), would replace total suspended particulate (TSP) as the basic particulate pollution parameter.

SUMMARY AND CONCLUSIONS

The broad outlines of deposition efficiency in the major functional regions of the human respiratory tract are reasonably well defined. There is a reliable body of consistent quantitative data for deposition in the head, TB and alveolar regions of healthy adults, and a reasonably clear understanding of the relationships between deposition of specific cytotoxic dusts in the alveolar

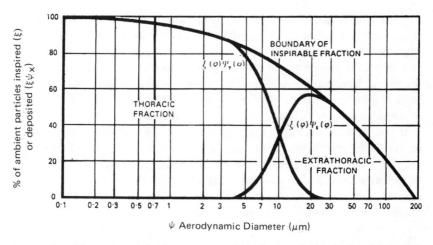

Figure 10.   Illustration of the figures proposed for the material inspired, and deposited in the extrathoracic or thoracic regions. The thoracic fraction is further subdivided into alveolar and TB fractions.

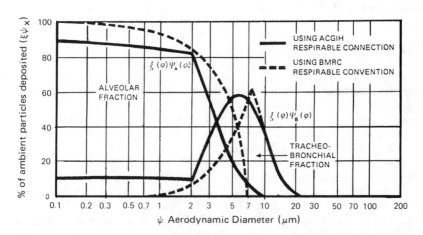

Figure 11.   Division of the thoracic fraction of Figure 10, into the alveolar and TB fractions where the target population is healthy adults. The operator may choose between the two conventions illustrated.

region and pathogenesis of pneumoconioses. Deposition of particles in the TB region has been associated with industrial bronchitis and bronchial cancers. Surface deposition is much higher at bifurcations of the larger bronchi than at other sites, as is the incidence of bronchial cancer. The bronchial surfaces associated with the pathogenesis of bronchitis are less well established.

Studies of the deposition efficiencies of inhaled particles in specific bronchial airways cannot be done in vivo. Therefore, intrabronchial distribution of particles has been studied as a function of particle size and electrical change and for various flow patterns, directions and rates, in hollow airway casts of the human larynx and upper bronchial airways. These studies, and some related studies in idealized physical models, provide useful insights into the relations between particle deposition and lung disease.

At this point, size-selective sampling criteria and recommendations reflecting available knowledge about regional particle deposition have been developed and applied by various national and international groups, and an up-to-date summary of these developments has been presented. None of these sampling criteria and recommendations have, as yet, addressed the issue of the intrabronchial variability of particle deposition.

ACKNOWLEDGMENTS

This study was supported by Grant ES 00881, and is part of a Center program supported by Grant ES 00260, both from the National Institute of Environmental Health Sciences.

REFERENCES

Altshuler, B., E. D. Palmes and N. Nelson (1967) "Regional Aerosol Deposition in the Human Respiratory Tract," in *Inhaled Particles and Vapours II*, C. N. Davies, Ed. (London: Pergamon Press, Ltd.), p. 323.

Armbruster, L., H. Breuer, D. Mark and J. H. Vincent (1982) "Definition and Measurement of Inhalable Dust," Chapter 17, this volume.

Bell, K. A., and S. K. Friedlander (1973) "Aerosol Deposition in Models of a Human Lung Bifurcation," *Staub-Reinhalt Luft* 33:178-182.

Chan, T. L., and M. Lippmann (1980) "Experimental Measurements and Empirical Modelling of the Regional Deposition of Inhaled Particles in Humans," *Am. Ind. Hygiene Assoc. J.* 41:399-409.

Chan, T. L., M. Lippmann, V. Cohen and R. B. Schlesinger (1978) "Effect of Electrostatic Charges on Particle Deposition in a Hollow Cast of the Human Larynx-Tracheobronchial Tree," *J. Aerosol Sci.* 9:463-468.

Chan, T. L., R. M. Schreck and M. Lippmann (1980) "Effects of the Laryngeal Jet on Particle Deposition in the Human Trachea and Upper Bronchial Airways," *J. Aerosol Sci.* 11:447-459.

EPA (undated) "Respiratory Deposition and Biological Fate of Inhaled Aerosols and SO_2," in *Air Quality Criteria for Particulate Matter and Sulfur Oxides, Vol. V* (Research Triangle Park, NC: Environmental Criteria and Assessment Office, U.S. EPA).

George, A. C., and A. J. Breslin (1967) "Deposition of Natural Radon Daughters in Human Subjects," *Health Phys.* 13:375-378.

Giacomelli-Maltoni, G., C. Melandri, V. Prodi and G. Tarroni (1972) "Deposition Efficiency of Monodisperse Particles in Human Respiratory Tract," *Am. Ind. Hygiene Assoc. J.* 33:603-610.

Gurman, J. L., R. B. Schlesinger and M. Lippmann (1980) "A Variable-Opening Mechanical Larynx for Use in Aerosol Deposition Studies," *Am. Ind. Hygiene Assoc. J.* 41:678-680.

Hounam, R. F., A. Black and M. Walsh (1969) "Deposition of Aerosol Particles in the Nasopharyngeal Region of the Human Respiratory Tract," *Nature* 221:1254-1255.

ISO (1981) "Size Definitions for Particle Sampling," International Standards Organization, *Am. Ind. Hygiene Assoc. J.* 42(5):A64-A68.

Lee, W., and C. S. Wang (1977) "Particle Deposition in Systems of Repeatedly Bifurcating Tubes," in *Inhaled Particles IV*, W. H. Walton Ed. (Oxford: Pergamon Press, Ltd.), pp. 49-59.

Lippmann, M. (1970) "Deposition and Clearance of Inhaled Particles in the Human Nose," *Ann. Otol. Rhinol. Laryngol.* 79:519-528.

Lippmann, M. (1977) "Regional Deposition of Particles in the Human Respiratory Tract," in *Handbook of Physiology*, D. H. K. Lee, H. L. Falk and S. D. Murphy, Eds. (Bethesda, MD: American Physiological Society), pp. 213-232.

Lippmann, M., and R. E. Albert (1967) "A Compact Electric-Motor Driven Spinning Disc Aerosol Generator," *Am. Ind. Hygiene Assoc. J.* 28:501-506.

Lippmann, M., and R. Albert (1969) "The Effect of Particle Size on the Regional Deposition of Inhaled Aerosols in the Human Respiratory Tract," *Am. Ind. Hyg. Assoc. J.* 30:257-275.

Lippmann, M., D. B. Yeates and R. E. Albert (1980) "Deposition, Retention and Clearance of Inhaled Particles," *Brit. J. Ind. Med.* 37:337-372.

Martens, A., and W. Jacobi (1973) "Die In-Vivo Bestimmung der Aerosolteilchendeposition im Atemtract bei Mund-Bzw. Nasenatmung," in *Aerosole in Physik, Medizin und Technik* (Bad Soden, FRG: Gesellschaft fur Aerosolforschung), pp. 117-121.

Martin, D. and W. Jacobi (1972) "Diffusion Deposition of Small-Sized Particles in the Bronchial Tree," *Health Phys.* 23:23-29.

Martonen, T. B., and J. Lowe (1982) "Assessment of Aerosol Deposition Patterns in Human Respiratory Tract Casts," Chapter 13, this volume.

Miller, F. J., D. E. Gardner, J. A. Graham, R. E. Lee, Jr., W. E. Wilson and J. D. Bachmann (1979) "Size Considerations for Establishing a Standard for Inhalable Particles," *J. Air Poll. Control Assoc.* 29:612-615.

Ogden, T. L. (1983) "Inhalable, Inspirable and Total Dust," Chapter 16, this volume.

Rudolf, G., and J. Heyder (1974) "Deposition of Aerosol Particles in the Human Nose," in *Aerosole in Naturwissenschaft, Medizin und Technik*, V. Böhlau, Ed. (Bad Soden, FRG: Gesellschaft für Aerosolforschung).

Schlesinger, R. B. (1980) "Particle Deposition in Model Systems of Human and Experimental Animal Airways," in *Generation of Aerosols and Facilities for Exposure Experiments*, K. Willeke, Ed. (Ann Arbor, MI: Ann Arbor Science Publishers, Inc.), pp. 553-575.

Schlesinger, R. B., and M. Lippmann (1972) "Particle Deposition in Casts of the Human Upper Tracheobronchial Tree," *Am. Ind. Hygiene Assoc. J.* 33:237-251.

Schlesinger, R. B., and M. Lippmann (1976) "Particle Deposition in the Trachea: *In vivo* and in Hollow Casts," *Thorax* 31:678-684.

Schlesinger, R. B., and M. Lippmann (1978) "Selective Particle Deposition and Bronchogenic Carcinoma," *Environ. Res.* 15:424-431.

Schlesinger, R. B., D. E. Bohning, T. L. Chan and M. Lippmann (1977) "Particle Deposition in a Hollow Cast of the Human Tracheobronchial Tree," *J. Aerosol Sci.* 8:429-445.

Schlesinger, R. B., J. Concato and M. Lippmann (1982) "Particle Deposition During Exhalation: A Study in Replicate Casts of the Human Upper Tracheobronchial Tree," Chapter 14, this volume.

Schlesinger, R. B., J. L. Gurman and M. Lippmann (in press) "Particle Deposition Within Bronchial Airways: Comparisons Using Constant and Cyclic Inspiratory Flows," in *Inhaled Particles V,* W. H. Walton, Ed. (London: Pergamon Press, Ltd.).

Shanty, F. (1974) "Deposition of Ultrafine Aerosols in the Respiratory Tract of Human Volunteers," PhD Dissertation, School of Hygiene and Public Health, Johns Hopkins University, Baltimore, MD.

Stahlhofen, W., J. Gebhart and J. Heyder (1980) "Experimental Determination of the Regional Deposition of Aerosol Particles in the Human Respiratory Tract," *Am. Ind. Hygiene Assoc. J.* 41:385-398.

Swift, D. L. (1982) "Aerosol Sampling Requirements to Assess the Health Risk of Special Particulates," Chapter 18, this volume.

Timbrell, V. (1972) "Inhalation and Biological Effects of Asbestos," in *Assessment of Airborne Particles,* T. T. Mercer, P. E. Mortow and W. Stober, Eds. (Springfield, IL: Charles C. Thomas, Publisher).

Weibel, E. R. (1963) *Morphometry of the Human Lung* (Berlin: Springer-Verlag).

Yu, C. P. (1978) "A Two Component Theory of Aerosol Deposition in Human Lung Airways," *Bull. Math. Biol.* 40:693-706.

Yu, C. P., and J. P. Hu (1982) "Diffusional Deposition of Ultrafine Particles in the Human Lung," Chapter 12, this volume.

CHAPTER 12

DIFFUSIONAL DEPOSITION OF ULTRAFINE PARTICLES IN THE HUMAN LUNG

C. P. Yu and J. P. Hu
Faculty of Engineering and Applied Sciences
State University of New York at Buffalo
Buffalo, New York

ABSTRACT

Diffusional deposition of ultrafine particles in the human lung during steady breathing and breath-holding is investigated in a model study. It is found that, for steady breathing, deposition fraction depends on the product of D and τ regardless of particle size, where D is the diffusion coefficient and τ is the breathing period. The flowrate in the respiratory cycle has only a minor effect on deposition. It is also shown that, under normal breathing conditions, particle deposition in the lung occurs solely by diffusion for particle diameter smaller than 0.1 μm, and that it is more effective to use these small particles for estimating the airway dimensions in a breath-holding experiment than particles in the sedimentation range.

INTRODUCTION

Particles suspended in the air continually undergo Brownian motion due to bombardment by air molecules. The mean displacement of a particle resulting from this motion is proportional to $(D\tau)^{1/2}$, where D is the particle diffusion coefficient and τ is the residence time. If the particles are initially placed in an enclosure with dimension ℓ, the fraction of particles lost to the wall due to Brownian motion is then a function of $(D\tau)^{1/2}/\ell$.

For particles with diameters smaller than 0.1 μm, Brownian motion is the

139

major process responsible for particle transport. When particles in this size range are inhaled into the respiratory system, it has been hypothesized that the fraction of inhaled particles lost to the airway surfaces will depend again on a single parameter $(D\tau)^{1/2}$, but now τ takes the value of a breathing period [Friedlander 1964; Heyder et al. 1980]. The physical situation here is actually more complex than diffusion in a single enclosure, because the lung consists of many generations of airways, each with different size, and the residence times for inhaled particles at different generations are therefore different.

At present, very little experimental data on lung deposition of ultrafine particles are available. Experimental data for 0.2-μm particles were obtained by Heyder et al. [1975]. For particles smaller than 0.1 μm, the only attempt to measure lung deposition was by Shanty [1974]. His particles, however, were hygroscopic, and the deposition data he obtained are difficult to interpret.

In this chapter, diffusional deposition of ultrafine particles is studied analytically, based on a deposition model developed earlier [Yu 1978; Yu and Diu in press; Yu and Thiagarajan 1979]. The model treats the airways as a one-dimensional time-dependent distributed system, in which the airway depth or generation number is used as the spatial variable. This formulism offers a simple mathematical solution for calculating deposition.

DEPOSITION FORMULAS

The deposition formulas for a single airway adopted in this study are the ones derived by Landahl [1963] and Ingham [1975]. They correspond to two idealized flow patterns, one for slug flow and the other for parabolic flow. The choice of the flow pattern in our model depends on the Reynolds number in the airway. For Reynolds numbers >2000, we assume a slug velocity profile to account for turbulent flow and the collection efficiency for an airway in the ith generation has the form [Landahl 1963]:

$$\eta_i = 4.0\Delta_i^{1/2}(1 - 0.444\Delta_i^{1/2} + ...) \tag{1}$$

For a parabolic velocity profile, we use [Ingham 1975]:

$$\eta_i = 1 - 0.819\exp(-14.63\Delta_i) - 0.0976\exp(-89.22\Delta_i) - 0.0325\exp(-228\Delta_i)$$
$$- 0.0509\exp(125.9\Delta_i^{2/3}) \tag{2}$$

Here

$$\Delta_i = D\ell_i/(d_i^2 u_i)$$

where ℓ_i = airway length
   d_i = airway diameter
   u_i = average flow velocity
   D = particle diffusion coefficient.

The value of D varies with particle size and is given, for example, by Friedlander [1977].

We define $t_i = \ell_i/u_i$, which is the time for the flow to pass through the airways of the ith generation. Then, for any prescribed breathing pattern, it is always possible to write t_i as a known fraction of τ such that $t_i = \alpha_i\tau$. Thus, $\Delta_i = (\alpha_i/d_i^2) (D\tau)$, η_i is a function of $D\tau$, and lung deposition calculated using Equations 1 and 2 will also be a function of $D\tau$ only. The implications of this result are that large and small particles may have the same deposition fraction if breathing periods are adjusted accordingly, and that for given particle size and breathing period, deposition fraction may not be the same if the breathing pattern in a cycle is different.

DEPOSITION RESULTS

Deposition calculations from our model have shown that under normal breathing conditions, diffusion alone accounts for more than 99% of total deposition produced by combined effects of diffusion, sedimentation and impaction for particle diameter smaller than 0.1 μm. Thus, for practical purposes, we may regard particles with diameters smaller than 0.1 μm as deposited by diffusion only. Figure 1 shows total deposition by diffusion at mouth breathing vs $D\tau$ for various lung models at 3000 cm^3 functional residual capacity (FRC) and 1000 cm^3 tidal volume. The breathing cycle consists of equal duration of inspiration and expiration with no respiratory pause. As we observed earlier, deposition is a function of $D\tau$ for a given lung model, regardless of particle size, although different lung models have different depositions. These differences arise from intersubject variability in lung geometry [Yu and Diu 1982]. The calculated depositions also agree fairly well with data of Heyder et al. [1975] for 0.2-μm particles at various breathing conditions, as shown in Table I.

If we let the time for the inspiratory phase of the respiratory cycle be τ_I and the time for expiration be τ_E, so that $\tau_I + \tau_E = \tau$, the influence of the flowrate on deposition can then be examined by varying the ratio τ_I/τ. Figures 2 and 3 present these results for Weibel's lung model, and they show that the flowrate has only a minor effect on deposition.

For a given lung morphology, larger lung volume corresponds to larger airway dimensions and therefore has smaller deposition. Figure 4 presents deposition vs $D\tau$ for three different initial lung volumes of a Weibel's lung. It is observed that lung volume has a relatively weak effect on tracheobronchial deposition. This is because increasing airway dimension also leads

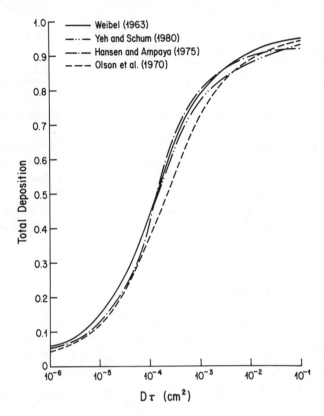

Figure 1. Total depositions vs $D\tau$ for 1000 cm^3 tidal volume, 15 breaths/min at FRC = 3000 cm^3. The breathing pattern consists of equal duration of inspiration and expiration with no pause. The data of various lung models are given elsewhere [Yu and Diu 1981].

to increasing residence time in this region, and the two effects on deposition tend to cancel each other. However, in the pulmonary region, considerably large deposition differences are found. For example, at $D\tau = 10^{-4}$ cm^2, an increase of lung volume from 3000 to 5000 cm^3, which is equivalent to an 18% increase in the linear dimensions of airways, will reduce pulmonary deposition by nearly half.

BREATH-HOLDING DEPOSITION

It has been proposed for many years [Lapp et al. 1975; Palmes 1973; Palmes et al. 1967] that aerosol deposition during breath-holding following

Table I. Comparison of Predicted Total Deposition with Experimental Data of Heyder et al. [1975] for 0.2-μm Particles at Nose and Mouth Breathing (Predicted Deposition was Obtained Using Weibel's Lung Model)

Subject	FRC (cm^3)	Tidal Volume (cm^3)	τ (sec)	Flowrate (cm^3/sec)	Deposition at Nose Breathing		Deposition at Mouth Breathing	
					Experimental	Predicted	Experimental	Predicted
1	2180	1000	8	250	0.25	0.26	0.20	0.23
3	2760	1000	8	250	0.31	0.25	0.21	0.22
4	4190	1000	8	250		0.22	0.23	0.18
5	4100	1000	8	250	0.24	0.22	0.22	0.18
6	3300	1000	8	250	0.26	0.24	0.22	0.21
1	2180	500	4	250	0.17	0.18	0.12	0.14
3	2760	500	4	250	0.22	0.16	0.13	0.12
4	4190	500	4	250		0.13	0.15	0.09
5	4100	500	4	250	0.17	0.13	0.15	0.09
6	3300	500	4	250	0.19	0.15	0.13	0.11
1	2180	1000	4	500	0.20	0.20	0.14	0.16
3	2760	1000	4	500	0.24	0.19	0.13	0.15
4	4190	1000	4	500		0.17	0.14	0.12
5	4100	1000	4	500	0.18	0.17	0.14	0.12
6	3300	1000	4	500	0.20	0.18	0.16	0.14
1	2180	1000	2	1000		0.16	0.08	0.11
3	2760	1000	2	1000		0.15	0.08	0.10
4	4190	1000	2	1000		0.14	0.08	0.08
5	4100	1000	2	1000	0.15	0.14	0.08	0.08
6	3300	1000	2	1000	0.15	0.15	0.09	0.09

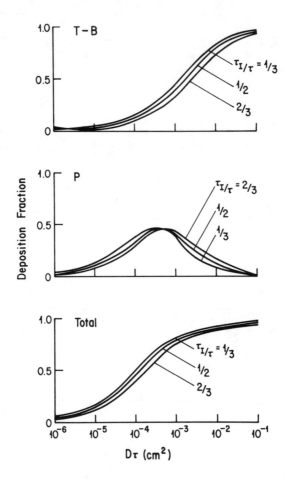

Figure 2.   Total and regional deposition vs $D\tau$ for several τ_I/τ at 1000 cm^3 tidal volume and 15 breaths/min. Weibel's lung model with FRC = 3000 cm^3 is used in the calculation.

inhalation of a single breath of aerosol can be used as a tool for estimating the size of airspaces in human lungs. Measurements and theoretical estimates of the fractional aerosol recovery show that it decays approximately in an exponential manner with the breath-holding time. However, these measurements and theories are mostly for particles in the sedimentation range [Beeckmans 1975; Heyder 1975; Landahl 1950; Yu and Thiagarajan 1979]. One obvious difficulty in dealing with sedimentational deposition in the breath-holding experiment is that the airways are normally inclined with

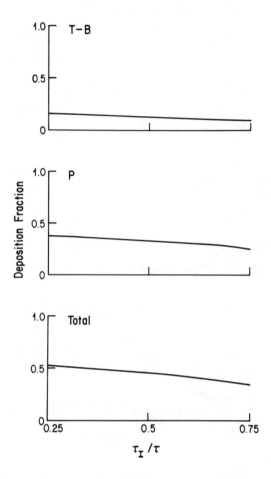

Figure 3.   Variations of total and regional deposition with τ_I/τ for $D\tau = 10^{-4}$ cm^2.

gravity and, because the inclination angles are unknown, it is difficult to assess exactly this orientation effect on lung deposition. If the airway orientation is assumed to be isotropically distributed in space, the average loss efficiency for an airway in the ith generation can be found to give [Yu and Thiagarajan 1979]:

$$\eta_i = 1.1094\epsilon_i - 0.2604\epsilon_i^2 \qquad \text{for} \quad 0 < \epsilon_i \leqslant 1 \qquad (3a)$$

$$\eta_i = 1 - 0.0069\epsilon_i^{-1} - 0.0859\epsilon_i^{-2} - 0.0582\epsilon_i^{-3} \quad \text{for} \quad \epsilon_i > 1 \qquad (3b)$$

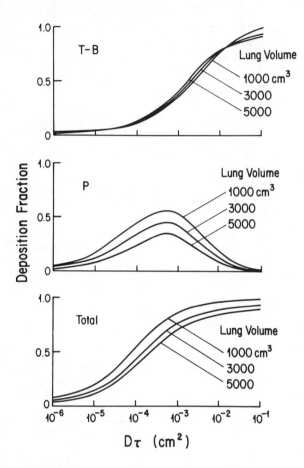

Figure 4. Total and regional deposition vs $D\tau$ at 1000 cm^3 tidal volume, 15 breaths/ min and various initial lung volumes. $\tau_I/\tau = 0.5$ is used in the breathing pattern.

Here,

$$\epsilon_i = u_s \tau_P / d_i$$

where u_s = particle settling velocity
        τ_P = breath-holding time

If particles in the diffusion range are used in the breath-holding experiment, deposition in an airway is independent of the airway orientation, and the collection efficiency may be written in the form [Ingham 1975]:

$$\eta_i = 1 - \sum_{j=1}^{3} \left\{ \frac{4}{k_j^2} \exp(-k_j^2 \Delta_i') - \left(1 - \sum_{j=1}^{3} \frac{4}{k_j^2}\right) \exp\left[-\frac{4\Delta_i'^{1/2}}{\pi^{1/2}\left(1 - \sum_{j=1}^{3} \frac{4}{k_j^2}\right)}\right] \right\} \quad (4)$$

where $\Delta_i' = D\tau p/d_i^2$

k_j = first three (j = 1, 2, and 3) roots of the equation $J_0(k) = 0$

Examination of Equations 3 and 4 reveals that for a given τ_P, η_i has a stronger dependence on d_i for the diffusional deposition. We may therefore conclude that breath-holding experiments attempting to measure the size of airspaces in the lung using inhaled particles will find that diffusional particles are more effective for this purpose. To further illustrate this point, Figure 5 shows the half-life of the aerosol recovery calculated for both 0.1- and 1.0-μm particles as a function of lung volume based on Weibel's lung model, where the half-life is defined as the time period in which the aerosol recovers half of its original amount after breath-holding. Since larger lung volume corresponds to greater airway sizes for a fixed lung structure, Figure 5 shows that the recovery of 0.1-μm particles has a stronger dependence on airway sizes than that of 1.0-μm particles.

CONCLUSIONS

Diffusional deposition of ultrafine particles in the human lung has been studied analytically. It has been concluded that, for steady breathing, the fraction of deposition depends on the product of $D\tau$ regardless of particle size. Calculated deposition of 0.2-μm particles agrees reasonably well with experimental data for several τ values. Because deposition experiments for very small particles are difficult to perform, this conclusion suggests that a larger particle size may be used for the experiment, provided that the value of τ is adjusted accordingly.

It has been also shown that diffusional deposition is more sensitive to airway sizes than is sedimentational deposition. Ultrafine particles are therefore better tools for measuring airway dimensions in a breath-holding experiment.

ACKNOWLEDGMENTS

This work was supported by Grant No. ES-02565 from the National Institute of Environmental Health Sciences and Grant No. OH-00923 from the National institute for Safety and Occupational Health.

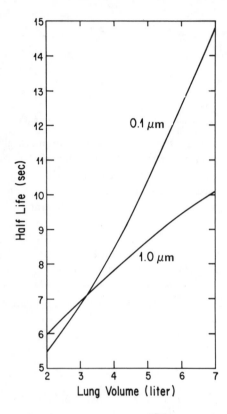

Figure 5. Half-life of the aerosol recovery as a function of the initial lung volume in a breath-holding experiment. The process consists of the inspiration of 1000 cm^3 aerosol in 2 sec followed by a breath-holding period and an expiration period of 4 sec.

REFERENCES

Beeckmans, J. M. (1975) "Analysis of Aerosol Decay During Breath-Holding," *Ann. Occup. Hygiene* 18:161.

Friedlander, S. K. (1964) "Particle Deposition by Diffusion in the Lower Lung: Application of Dimensional Analysis," *Ind. Hygiene J.* Vol. 37.

Friedlander, S. K. (1977) *Smoke, Dust and Haze* (New York: John Wiley & Sons, Inc.).

Hansen, J. E., and E. P. Ampaya (1975) "Human Air Space, Shapes, Sizes, Areas and Volumes," *J. Appl. Physiol.* 38:990.

Heyder, J. (1975) "Gravitational Deposition of Aerosol Particles Within a System of Randomly Oriented Tubes," *J. Aerosol Sci.* 6:133.

Heyder, J., L. Armbruster, J. Gebhart, E. Grien and W. Stahlhofen (1975) "Total Deposition of Aerosol Particles in the Human Respiratory Tract for Nose and Mouth Breathing," *J. Aerosol Sci.* 6:311.

Heyder, J., J. Gebhart, G. Rudolf and W. Stahlhofen (1980) "Physical Factors Determining Particle Deposition in the Human Respiratory Tract," *J. Aerosol Sci.* 11:505.

Ingham, D. B. (1975) "Diffusion of Aerosols from a Stream Flowing Through a Cylindrical Tube," *J. Aerosol Sci.* 6:125.

Landahl, H. D. (1950) "On the Removal of Airborne Droplets by the Human Respiratory Tract: I. The Lung," *Bull. Math. Biophys.* 12:43.

Landahl, H. D. (1963) "Particle Removal by the Respiratory System—Note on the Removal of Airborne Particulates by the Human Respiratory Tract with Particular Reference to the Role of Diffusion," *Bull. Math. Biophys.* 25:29.

Lapp, N. L., J. L. Hankinson, H. Amandus and E. D. Palmes (1975) "Variability in the Size of Airspaces in Normal Human Lungs as Estimated by Aerosols," *Thorax* 30:293.

Olson, D. E., G. A. Dart and G. F. Filley (1970) "Pressure Drop and Fluid Flow Regime of Air Inspired into the Human Lung," *J. Appl. Physiol.* 28:482.

Palmes, E. D. (1973) "Measurement of Pulmonary Air Spaces Using Aerosols," *Arch. Intern. Med.* 131:76.

Palmes, E. D., B. Altshuler and N. Nelson (1967) "Deposition of Aerosols in the Human Respiratory Tract During Breath Holding," in *Inhaled Particles and Vapours, Vol. 2,* C. N. Davies, Ed. (Oxford: Pergamon Press, Ltd.), p. 339.

Shanty, F. (1974) "Deposition of Ultrafine Aerosols in the Respiratory Tract of Human Volunteers," PhD Dissertation, Johns Hopkins University.

Weibel, E. R. (1963) *Morphometry of Human Lung* (New York: Academic Press).

Yeh, H. C., and G. M. Schum (1980) "Models of Human Lung Airways and Their Applications to Inhaled Particle Deposition," *Bull. Math. Biol.* 42:461.

Yu, C. P. (1978) "Exact Analysis of Aerosol Deposition During Steady Breathing," *Power Technol.* 21:55.

Yu, C. P., and C. K. Diu (1982) "A Comparative Study of Aerosol Deposition in Different Lung Models," *Am. Ind. Hygiene Assoc. J.* 43:54.

Yu, C. P., and V. Thiagarajan (1979) "Decay of Aerosols in the Lung During Breath Holding," *J. Aerosol Sci.* 10:11.

CHAPTER 13

ASSESSMENT OF AEROSOL DEPOSITION PATTERNS IN HUMAN RESPIRATORY TRACT CASTS

T. B. Martonen* and J. Lowe

Pulmonary Division
Department of Medicine
University of California
Irvine, California

ABSTRACT

To assess the human health hazard presented by airborne particulate matter in the mining and industrial work environment, information is needed concerning total dose deposition and distribution. Data have been obtained by depositing monodisperse ammonium fluorescein aerosols in respiratory system simulators—combined human replica larynx casts and single-pathway tracheo-bronchial (TB) tree models. The latter, which were custom-made, followed Weibel's morphology. Since they have only two airways in each generation distal to the trachea, air flowrates and patterns could be controlled in a practical manner with rotameters. Larynx configurations corresponded to inspiratory flowrates of 15, 30 and 60 liter/min. The mass median aerodynamic diameters (MMAD) of the aerosols ranged from 1.9 to 10.6 μm, with geometric standard deviations of 1.11–1.17. Total larynx and TB deposition measurements could be expressed in terms of a single parameter, the particle Stokes number, indicating the significance of inertial impaction. The larynx casts and some airway bifurcations were sites of enhanced deposition. Such "hot spots" indicate very high dosage to epithelial cells of workers' airways, and may have important implications regarding the establishment of threshold exposure limits.

---

*Present address: Northrop Services, Inc., Research Triangle Park, NC.

INTRODUCTION

Assessment of the health hazards presented by airborne particulate matter in the mining and industrial work environment requires knowledge concerning deposition sites within the respiratory tract. Valuable information about dose distribution can be obtained from inhalation exposure tests using human subjects. Such tests, however, yield few data about sites of enhanced deposition. Schlesinger et al. [1977] demonstrated the usefulness of conducting deposition studies with models of the human respiratory system consisting of combined replica casts of the larynx and TB tree. The in vitro experimental data were shown to be in good agreement with findings from in vivo human test exposures. However, the uniqueness of the replica TB cast used in the study precluded cutting it apart to study localized deposition patterns. Therefore, a specially devised laboratory system consisting of three multiple-channel focusing collimators for use with a NaI scintillation crystal was required to quantify aerosol deposition efficiencies.

We have incorporated replica larynx casts with hand-made TB casts for use in studying the effects in humans of particle size and inspiratory flowrate on aerosol behavior. The casts, which can be produced in unlimited numbers, can therefore be dissected to measure the aerosol doses delivered to specific sites. Moreover, doses can be measured using common fluorometric procedures. These morphological models are used here to quantify particle deposition efficiencies in the upper TB tree, and to measure localized deposits occurring in the larynx and at airway branching sites.

MORPHOLOGICAL MODELS

Three Silastic® E RTV silicone rubber casts were molded from three original epoxy resin modeling clay casts of human larynges obtained at autopsy [Schlesinger et al. 1977]. The internal geometries of the individual laryngeal models were related to inspiratory air flowrates of 15, 30 and 60 liter/min through them. The interiors of the original models were coated with a releasing agent before being filled with the silicone rubber to make the casts described here. After curing, the solid casts were simply pulled out. They then became solid "master" casts for making hollow replicas, after being painted with a releasing agent and coated with the silicone rubber. Once the rubber layer hardened, the solid "master" was removed. The resultant hollow larynx casts had internal geometries that were exact copies of the original cadaveric casts. Axial-view photographs of the casts used in this work are given in Figure 1. The cross-sectional areas of the glottis openings are 0.88, 1.40 and 2.42 cm^2, respectively.

Hollow TB casts were fabricated from the Silastic E RTV silicone rubber

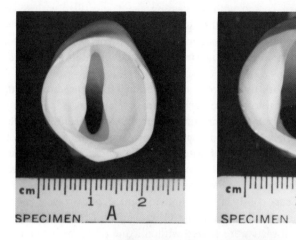

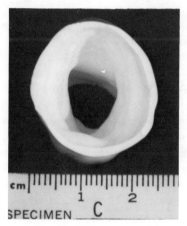

Figure 1.   Internal configurations of human larynx casts. The glottis openings correspond to steady inspiratory flow rates of (A) 15 (B) 30 and (C) 60 liter/min.

material using the "lost wax" process. A solid wax model was hand-made,
following the length and diameter dimensions of a symmetric, dichotomous
branching pattern proposed by Weibel [1963]. A branching angle of 70°
between airways of the same generation was selected to be compatible with
anatomical data [Horsfield et al 1971]. Silicone rubber was painted onto the
wax model and allowed to cure. The composite was subsequently heated, and
the wax core was poured out. The interior of this hollow "master" silicone

rubber cast was then coated with a releasing agent and filled with Silastic. When cured, the solid core was pulled out of the surrounding shell. The resultant solid cast was a faithful reproduction of the original solid wax cast and subsequently served as the solid "master" for producing hollow TB casts for experiments.

The branching network used here to describe the TB tree produces a single-pathway TB cast, consisting of a trachea and two airways in each of five generations of distal bronchi. The airways downstream from the trachea correspond to the main, lobar and segmental bronchi of the human. In this pattern, branching to lobar bronchi distal to the main bronchi continues from the left main bronchus. The right member of that generation then branches into segmental airways of the next generation. This left-right alternating scheme results in a cast consisting of a limited number of airways, which has some practical advantage, discussed later in this chapter. The smooth interiors of the larynx and TB casts were coated with silicone oil immediately before experiments to simulate the natural mucus lining of the upper respiratory tract of the human. This served to limit particle bounce-off and possible redistribution of deposited aerosol during tests.

EXPERIMENTAL SYSTEM

Monodisperse fluorescein aerosols were generated with a spinning-disc instrument. Fluorescein aerosols were utilized because minute quantities of deposited mass could easily be measured using standard fluorometric techniques and detection equipment. The ammonium salt of fluorescein was used because it is nonhygroscopic [Stöber and Flachbart 1973]. Uranine, the sodium salt of fluorescein, has been used in aerosol deposition studies by Scherer et al. [1979] and Johnson and Schroter [1979]. However, uranine is hygroscopic, a property that may lead to the misinterpretation of laboratory findings unless relative humidity conditions during experiments are accurately controlled to ensure that particle growth by water vapor absorption does not occur [Knight and Tillery 1967].

Characteristics of the particle size distributions of the aerosols generated are presented in Table I. The droplets produced by the spinning disc unit were dried to solid particles by passing the aerosol through an annular heating coil encased in a ceramic housing. These particles then entered an electric charge neutralizer, where they were introduced to a high concentration of bipolar ions produced by a sealed 20-mCi source of ^{85}Kr. The final electric charge distribution of an aerosol traveling through such a neutralizer closely approximates the equilibrium Boltzmann distribution.

The laboratory system used to regulate the flow of aerosol through the joined larynx and TB casts is illustrated in Figure 2. The flowrate through

Table I. Conditions for Aerosol Deposition Experiments

Aerosol Parameters		Inspiratory Flowrate,
MMAD (μm)	σ_g[a]	Q_0 (liter/min)
1.9	1.17	15
3.0	1.12	15
6.8	1.11	30
8.7	1.16	60
10.6	1.14	60

[a]Geometric standard deviation.

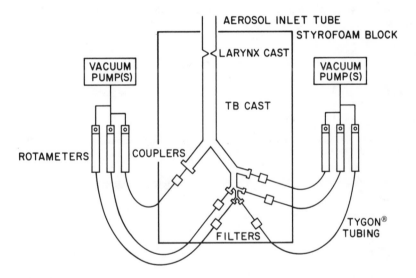

Figure 2.   Experimental system to control aerosol flow distribution through the combined larynx and tracheobronchial casts.

each TB airway was controlled by its own rotameter. The limited number of airways in the single-pathway TB casts permitted the flowrates to be regulated efficiently and accurately in this practical manner. At every branching location, airstreams were divided symmetrically. Experiments were conducted at inspiratory flowrates, Q_0, of 15, 30 and 60 liter/min with the appropriate larynx cast in place.

The larynx casts were included in the system for two reasons: (1) to ensure a physiologically suitable airflow pattern simulating the laryngeal jet in the TB casts; and (2) to measure particle losses in the larynx. The laryngeal jet, generated at the restricted glottis opening, has been demonstrated to affect particle deposition efficiencies in the trachea [Schlesinger and Lippmann 1976]. Unfortunately, the influence of the laryngeal jet on particle motion has been ignored in some recent aerosol deposition studies with TB models, which suggests that those findings may not be applicable to human upper airways [Johnson and Schroter 1979; Scherer et al. 1979].

EXPERIMENTAL RESULTS

Aerosol deposition efficiencies achieved in our experiments within the larynx casts are plotted in Figure 3. The particle Stokes number is defined as:

$$Stk = \rho D^2 V / 18 \mu R \; ,$$

where ρ = particle density
D = particle geometric diameter
μ = air viscosity
V = mean airstream velocity
R = 1.65 cm (this was the largest laryngeal opening measured at all flowrates)

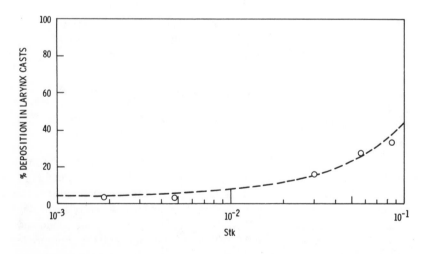

Figure 3.   Percent of aerosol mass entering larynx casts that is deposited as a function of particle Stokes number, Stk.

The dashed line in Figure 3 is an empirical laryngeal deposition probability curve P_Q based on linear regression analysis of the experimental data presented here with data from a separate study using the same experimental system [Martonen in press]. The latter set of data will be reported in another publication. For $10^{-3} < Stk < 10^{-1}$, the curve is described by the formula:

$$P_Q = 3.50 + 390 \, Stk \qquad (1)$$

with a correlation of $r = 0.96$. Equation 1 agrees well with another formula used to describe larynx losses when $10^{-3} < Stk < 2 \times 10^{-2}$ [Chan et al. 1978]. However, for Stk values in the $2 \times 10^{-2} < Stk < 10^{-1}$ interval, Equation 1 predicts lower larynx losses.

The data indicate that inertial impaction is the dominant deposition mechanism affecting particle behavior in the larynx since P_Q can be expressed accurately as a function of a single parameter, Stk. Equation 1 may be used to compute in vivo losses in the human larynx during inspiration. The formula could also be used to estimate losses occurring during the expiration phase of a breathing cycle, since it has been demonstrated that aerosol deposition efficiencies within another region of the human upper respiratory tract, the complex nasopharyngeal compartment, are essentially the same during the inspiration and expiration phases of in vivo experiments [Heyder and Rudolf 1977].

Data related to the surface dose, L_d, deposited within the larynx casts are given in Table II. The value of L_d is calculated from

$L_d = $ [(aerosol mass deposited in the larynx casts/surface area of the larynx casts)

$\div$ (total aerosol mass deposited in the combined larynx and TB casts/

total surface area of the combined larynx and TB casts)] $\times$ 100

Table II. Measurement of the Degree of Enhanced
Aerosol Deposition Within the Larynx, L_d[a]

Inspiratory Flowrate, Q_0 (liter/min)	MMAD (μm)	Stk[b]	L_d (%)
15	1.9	1.9×10^{-3}	410
15	3.0	4.8×10^{-3}	390
30	6.8	3.0×10^{-2}	430
60	8.7	5.6×10^{-2}	380
60	10.6	8.4×10^{-2}	330

[a]Data are normalized to a uniform distribution of aerosol throughout the combined larynx-TB casts.
[b]Particle Stokes number.

Thus, L_d is the percent of enhanced deposition that occurs within the larynx casts when compared to a uniform distribution of deposited material throughout the combined larynx-TB models of the upper respiratory tract. Therefore, for example, the value of 330% at Stk = 8.4×10^{-2} indicates that the larynx dose is a factor of 3.3 greater than if the aerosol had been uniformly dispersed.

Total aerosol deposition efficiencies within the single-pathway TB casts are shown in Figure 4 as a function of Stk. In computing Stk values pertinent for the TB tree, the following parameters were used: R = radius of the trachea (0.9 cm) and V = mean airstream velocity in the trachea. The data indicate that inertial impaction is also the most effective mechanism of deposition in large bronchial airways. The plotted data follow a pattern similar to that found in aerosol deposition experiments by Schlesinger and Lippmann [1972] with a replica cast of the human TB tree that was pruned to the segmental bronchi.

The pattern of aerosol deposition at branching sites B_d is shown in Table III. The tabulated values for the bifurcation dose at each generation are:

B_d = [(aerosol mass deposited within a bifurcation/bifurcation surface area)

   $\div$ (total aerosol mass deposited within two airways of a generation,

     including their common bifurcation/total airway and

     bifurcation surface area)] $\times$ 100

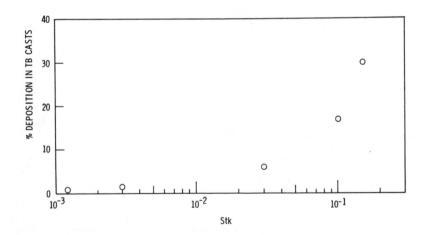

Figure 4.   Percent of aerosol mass entering the trachea that is deposited within the tracheobronchial casts as a function of particle Stokes number, Stk.

Table III. Percent Efficiency of Aerosol Deposition at Airway Bifurcations[a]

TB Cast Generation	Q_0 = 15 liter/min		Q_0 = 30 liter/min	Q_0 = 60 liter/min	
	MMAD = 1.9 μm	MMAD = 3.0 μm	MMAD = 6.8 μm	MMAD = 8.7 μm	MMAD = 10.6 μm
1	80	51	65	101	120
2	115	115	135	140	152
3	150	162	231	305	316
4	91	87	157	210	194
5	110	91	150	175	160

[a]Data are normalized to a uniform distribution of deposited particles within a generation

Therefore, the data represent the extent to which particulate matter collected within a given generation is preferentially deposited at bifurcation sites. In each experiment, the bifurcation that received the maximum dose was located at airway generation 3. Indeed, for the 10.6-μm-MMAD aerosol and an inspiratory flowrate Q_0 = 60 liter/min, deposition was enhanced by a factor of 3.16. This means that epithelial cells located at that bifurcation in the human TB tree may receive more than three times the dose of other airway cells in that generation. In every experiment, the bifurcation at the branching site between the trachea and main bronchi received the lowest dose of all bifurcations. This can be attributed to the relatively large bifurcation surface area at that location.

The distribution of particulate matter within the individual bifurcation regions was not uniform. Visual inspection of those dissected sections of the TB casts revealed highly concentrated deposits at the carina-ridge of each bifurcation. These observations are consistent with experimental findings that measured local nonuniformities in deposition patterns in an idealized model of the human trachea and main bronchi that was machined from an aluminum block [Bell and Friedlander 1973]. In that work, airway walls were lined with adhesive tape, which could be removed to directly measure local particle transfer coefficients using optical techniques. "Hot spots" (peaks of deposition) were detected at the carina itself and at sites slightly downstream along the inner wall of the main bronchi. Lippmann and Altschuler [1976] also detected concentrated deposits at dividing "spurs" at each airway branching site within replica TB casts. The most dense surface deposits were found at lobar and segmental bronchi locations. Chan et al. [1980] suggested that localized airflow patterns may be the most significant factors affecting particle deposition probabilities at such sites.

DISCUSSION

Data from in vitro aerosol deposition experiments indicate that certain regions in casts of the human upper respiratory tract receive higher doses than others. The findings suggest that the larynx and bifurcation sites within the TB tree may be locations where inhaled aerosols are preferentially deposited in vivo following exposure to airborne contaminants. The higher surface density of deposited material due to such a dose distribution pattern increases the potential of adverse health effects being initiated at epithelial cells within the highly localized deposits. The doses delivered to such "hot spots" should be accounted for in evaluating the health hazards of aerosols in the mining and industrial work environment.

It has been established clearly [Garland et al. 1962; Schlesinger et al. 1977; Veeze 1968] that the efficiencies with which particles are initially deposited in certain airways in TB casts correlate with their frequency of involvement as sites of primary bronchial carcinoma in the TB tree in vivo. The highest incidence of primary cancer in all airways occurs in main to segmental bronchi. The actual distribution of bronchogenic cancer within those large airways is preferential at lobar and immediately distal airways (generations I = 2 and 3), exists to a lesser degree in segmental bronchi (I = 4 and 5), and is found least often in main bronchi (I = 1).

There is experimental evidence that bronchogenic carcinomas actually originate at bifurcation sites in the human TB tree [Auerbach et al. 1961; Ermala and Holsti 1955; Kotin and Falk 1959]. Table III indicates the extent to which aerosol particles may be deposited at such branching sites in vivo. Comparable data, using replica larynx-TB casts, have been presented by Schlesinger et al. [1977]. In vivo, the high surface concentrations of hazardous aerosols at bifurcations may not be due to direct deposition alone. Epithelial cells situated there have increased exposures relative to other airway cells because the mucociliary blanket has reduced movement at branching locations [Hilding 1957]. Furthermore, Lippmann and Altschuler [1976] have reported that common airborne contaminants can significantly affect the function of the mucociliary clearance mechanism. Their findings suggest that, under adverse air conditions, all TB epithelial cells may have increased exposure to toxic constituents of inhaled particles because of the slowed function of the clearance mechanism. This effect will be exacerbated within the lungs of tobacco smokers, because the smoke itself has been demonstrated [Albert et al. 1971] to reduce mucociliary clearance in the human, due either to decreased ciliary activity or changes in properties of the mucus sheath.

Experimental evidence indicates that laboratory animals used as models to evaluate the toxicity of aerosols to which humans are exposed also may have greater deposition in the larynx and trachea than in other parts of the respiratory tract. Inhalation exposure of hamsters to tobacco smoke has demonstrated that preneoplastic and neoplastic lesions can be produced in the larynx without being observed elsewhere in the respiratory tract [Bernfeld et al, 1974; Dontenwill et al. 1973]. Epithelial hyperplasia has been detected in the larynx of the rat following exposure to tobacco smoke [Davis et al. 1975]. Furthermore, squamous metaplasia was observed in the larynx of each rat of a colony in a chronic tobacco smoke exposure investigation [Dalbey et al. 1980]. In more recent acute studies, Burton et al. [in press] exposed rats to various red phosphorus atmospheres. Pathological examination of tissue from the upper and lower respiratory tract revealed that, in all exposure groups, lesions most consistently occurred in the epiglottal and tracheal regions. Also, tracheal lesions consisting of fibrin-like coatings of mucosa immediately distal to the larynx were found at the highest aerosol concentrations. The tracheal lesions may have been "hot spots" of deposition caused by particles entrained in the laryngeal jet being directed against tracheal surfaces downstream from the glottis opening.

The TB casts used in this study simulate the large ciliated airways of the human. Particulate matter deposited in those airways in vivo will, of course, be subject to clearance by the mucociliary system. To assess accurately the potential of adverse health effects, the rate of clearance and redistribution of the deposited aerosol should be ascertained. If particles are soluble, toxic components may be leached out quickly, damaging epithelial cells. Whether this occurs in cells lining initial deposition sites, or in other cells following redistribution, depends on the relationship of clearance rates vs solubility of particle constituents. For insoluble particles, the high doses measured at the localized sites of enhanced deposition will have particular significance to toxic particles deposited and retained there. This subject of particle uptake by large-airway epithelium is being investigated because it is a major site of carcinomas following human exposures to radioactive particles and environmental containments [Patrick 1979; Watson and Brain 1979]. Therefore, the most important applications of the method for quantification of aerosol deposition reported here may be in determining the hazards to human health presented by particles highly soluble in respiratory tract fluids as well as those of insoluble particles whose toxic effects are tissue-specific.

ACKNOWLEDGMENTS

This work was conducted with the financial support of National Institutes of Health Grant No. NHLBI-R01-HL-19704 and Battelle Pacific Northwest Laboratories Special Study No. WBS 16-02.

REFERENCES

Albert, R. E., M. Lippman and H. T. Peterson, Jr. (1971) "The Effects of Cigarette Smoking on the Kinetics of Bronchial Clearance in Humans and Donkeys," in *Inhaled Particles III* W. H. Walton, Ed. (Old Woking, UK: Unwin Bros, Ltd., The Gresham Press).

Auerbach, O., A. P. Stout, E. C. Hammond and L. Garfinkel (1961) "Changes in Bronchial Epithelium in Relation to Cigarette Smoking and in Relation to Lung Cancer," *New England J. Med.* 265:253.

Bell, K. A., and S. K. Friedlander (1973) "Aerosol Deposition in Models of a Human Lung Bifurcation," *Staub-Reinhalt Luft* 33:183.

Bernfeld, P., F. Homburger and A. B. Russfeld (1974) "Strain Differences in the Response of Inbred Syrian Hamsters to Cigarette Smoke Inhalation," *J. Nat. Cancer Inst.* 53:1141.

Burton, F. G., M. L. Clark, R. A. Miller and R. E. Schirmer (in press) "Generation and Characterization of Red Phosphorus Smoke Aerosols for Inhalation Exposure of Laboratory Animals," *Am. Ind. Hygiene Assoc. J.*

Chan, T. L., M. Lippmann, V. R. Cohen and R. B. Schlesinger (1978) "Effect of Electrostatic Charges on Particle Deposition in a Hollow Cast of the Human Larynx-Tracheobronchial Tree," *J. Aerosol Sci.* 9:463.

Chan, T. L., R. M. Schreck and M. Lippmann (1980) "Effect of the Laryngeal Jet on Particle Deposition in the Human Trachea and Upper Bronchial Airways," *J. Aerosol. Sci.* 11:447.

Dalbey, W. E., P. Nettesheim, R. Griesemer, J. E. Canton and M. R. Guerin (1980) "Lifetime Exposures of Rats to Cigarette Tobacco Smoke," *Pulmonary Toxicology of Respirable Particles*, C. L. Sanders, F. T. Cross, G. E. Dagle and J. A. Mahaffey, Eds. CONF-791002 (Springfield, VA: NTIS).

Davis, B. R., J. K. Whithead, M. E. Gill, P. N. Lee, A. D. Butterworth and F. J. C. Roe (1975) "Response of Rat Lung to Inhaled Tobacco Smoke with or without Prior Exposure to 3,4-Benzopyrene (BP) Given by Intratracheal Instillation," *Brit. J. Cancer* 31:469.

Dontenwill, W., H. H. Chevalier, H. P. Harke, V. Lafrenz, G. Reckzh and B. Schneider (1973) "Investigations on the Effects of Chronic Cigarette-Smoke Inhalation in Syrian Golden Hamsters," *J. Nat. Cancer Inst.* 51:1781.

Ermala, P., and L. R. Holsti (1955) "Distribution and Absorption of Tobacco Tar on the Organs of the Respiratory Tract," *Cancer* 8:673.

Garland, L. H., R. L. Beier, W. Carlson, J. H. Heald and R. L. Stein (1962) "The Apparent Sites of Origin of Carcinomas of the Lung," *Radiology* 78:1.

Heyder, J., and G. Rudolf (1977) "Deposition of Aerosol Particles in the Human Nose," in *Inhaled Particles IV*, W. H. Walton, Ed. (Oxford: Pergamon Press).

Hilding, A. C. (1957) "Ciliary Streaming in the Bronchial Tree and the Time Element in Carcinogens," *New England J. Med.* 256:634.

Horsfield, K., G. Dart, D. E. Olson, G. F. Filley and J. Cummings (1971) "Models of the Human Bronchial Tree," *J. Appl. Physiol.* 31:207.

Johnson, J. R., and R. L. Schroter (1979) "Deposition of Particles in Model Airways," *J. Appl Physiol.* 47:947.

Knight, M. E., and M. I. Tillery (1967) "On the Density of Uranine," *Am. Ind. Hygiene Assoc. J.* 28:498.

Kotin, P., and H. L. Falk (1959) "The Role and Action of Environmental Agents in the Pathogenesis of Lung Cancer, I. Air Pollutants," *Cancer* 12:147.

Lippmann, M., and B. Altschuler (1976) "Regional Deposition of Aerosols," in *Air Pollution and the Lung*, E. F. Aharonson, A. Ben-David and M. A. Klingberg, Eds (New York: John Wiley & Sons, Inc.).

Martonen, T. B. (in press) "Measurement of Particle Dose Distribution in a Model of a Human Larynx and Tracheobronchial Tree," *J. Aerosol Sci.*

Patrick, G. (1979) "The Retention of Uranium Dioxide Particles in the Trachea of the Rat," *Int. J. Rad. Biol.* 35:571.

Scherer, P. W., F. R. Haselton, L. M. Hanna and D. R. Stone (1979) "Growth of Hygroscopic Aerosols in a Model of Bronchial Airways," *J. Appl. Physiol.* 47:544.

Schlesinger, R. B., and M. Lippmann (1972) "Particle Deposition in Casts of the Human Upper Tracheobronchial Tree," *Am. Ind. Hygiene Assoc. J.* 33:237.

Schlesinger, R. B., and M. Lippmann (1976) "Particle Deposition in the Trachea: *In Vivo* and in Hollow Casts," *Thorax* 31:678.

Schlesinger, R. B., D. E. Bohning, T. L. Chan and M. Lippmann (1977) "Particle Deposition in a Hollow Cast of the Human Tracheobronchial Tree," *J. Aerosol Sci.* 8:429.

Stöber, W., and H. Flaschbart (1973) "An Evaluation of Nebulized Ammonium Fluorescein as a Laboratory Aerosol," *Atmos. Environ.* 7:737.

Veeze, P. (1968) *Rationale and Methods of Early Detection in Lung Cancer* (Assen: Van Gorcum and Co.).

Watson, A. Y., and J. D. Brain (1979) "Uptake of Iron Oxide Aerosols by Mouse Airway Epithelium," *Lab. Invest.* 40:450.

Weibel, E. (1963) *Morphometry of the Human Lung* (Berlin: Springer-Verlag).

CHAPTER 14

PARTICLE DEPOSITION DURING EXHALATION: A STUDY IN REPLICATE CASTS OF THE HUMAN UPPER TRACHEOBRONCHIAL TREE

Richard B. Schlesinger, John Concato and Morton Lippmann

Institute of Environmental Medicine
New York University Medical Center
New York, New York

ABSTRACT

Deposition of particles in the human upper tracheobronchial (TB) tree during exhalation was studied using hollow airway casts. Exposures were performed at constant flowrates of 15 and 30 liter/min with monodisperse ^{99m}Tc-tagged ferric oxide aerosols having mass median aerodynamic diameters (MMAD) ranging ~3–6 μm. Two replicate RTV silicone casts connected at distal ends with latex tubing were used for each exposure. Exhalation deposition efficiency was found to be a significant fraction of deposition efficiency during inhalation. Bifurcations often were sites of preferential deposition under exhalation as well as under inhalation. Particle deposition during exhalation cannot be ignored when considering aerosol deposition in the TB tree.

INTRODUCTION

The regional pattern of deposition within the respiratory tract is an important factor in determining the potential toxicologic hazard from inhaled particles. For example, the pathogenesis of some lung diseases, e.g., chronic bronchitis and bronchogenic carcinoma, involves deposition of aerosols within the bronchial tree. Thus, information concerning TB deposition is essential for analysis of risk from many inhaled aerosols.

165

The interactions of the factors that determine particle deposition within the TB tree are complex. Thus, mathematical models, ideal physical models and hollow airway casts have been used to aid in experimental analyses, usually with the ultimate goal being the formulation of predictive deposition equations for risk estimates [Schlesinger 1980].

Although a complete breathing cycle consists of two phases, i.e., inhalation and exhalation, most of the studies employing model systems examined only the former phase. Thus, there are very few experimental data concerning TB deposition patterns on expiration, an essential component for any complete dosimetric model. Those data that exist are for particles less than ~2 μm. Martin and Jacobi [1972] studied deposition of submicron radioaerosols during inhalation and exhalation using a plastic model of the human upper tracheobronchial tree having dimensions of and extending to the first four generations of the Weibel Model "A" [Weibel 1963]. Wang and Son [1977] studied the deposition of 1.14- and 2.05-μm (aerodynamic diameter) particles during exhalation in a symmetrical glass model consisting of four successive generations of bifurcations.

Mathematical deposition models for spherical particles generally account for both breath phases, but make certain assumptions about differences in deposition mechanisms between them, e.g., during expiration, deposition due to impaction is zero, and particle removal is due only to sedimentation and diffusion [Austin et al. 1979; Beeckmans 1965; Landahl 1950; Findeisen 1935; Schum and Yeh 1980; Yu and Taulbee 1977].

This chapter describes a study of the deposition of ~3- to 6-μm-MMAD particles during constant inhalation and exhalation using replicate hollow casts of the human upper TB tree, which extended through the fourth branching generation. Total deposition for each generation was measured, as were local patterns of intrabronchial deposition.

EXPERIMENTAL PROCEDURE

Production of Replicate Hollow Tracheobronchial Tree Casts

The method of production of replicate, hollow RTV silicone rubber casts from a solid RTV silicone master cast has been described [Schlesinger et al. in press]. Eight replicate casts were made for the experiment reported here. Each extended from branching generation 0 (trachea) through 4. A summary of cast morphometry is presented in Table I.

Aerosol Generation and Sizing

The test aerosols consisted of monodisperse ferric oxide microspheres tagged with the gamma-emitting radioisotope ^{99m}Tc, and produced by a

Table I. Cast Morphometry

Airway Generation	Number of Airways	Diameter (cm), Mean ± Standard Deviation	Length (cm), Mean ± Standard Deviation	Angle of Branching (degrees), Mean ± Standard Deviation	Angle from Horizontal (degrees), Mean ± Standard Deviation
0	1	1.45	9.20		
1	2	1.08 ± 0.10	3.65 ± 1.56	29 ± 3	61 ± 3
2	4	0.88 ± 0.10	1.24 ± 0.50	44 ± 31	44 ± 42
3	9	0.60 ± 0.14	0.91 ± 0.42	43 ± 20	46 ± 27
4	18	0.42 ± 0.10	0.89 ± 0.47	31 ± 16	44 ± 27

spinning-disc generator [Lippmann and Albert 1967]. The aerosols were charge-neutralized by passing the stream past a stainless steel capillary tube containing 20 mCi of ^{85}Kr.

The particle size distribution was determined after collection on a 25-mm-diameter membrane filter (Millipore®, Type AA) during each test run. The diameters of 50 particles per test sample were measured with a 100X oil-immersion objective and filar micrometer eyepiece. Using the technique of Hatch and Choate [1929], measured diameters were then converted to mass median diameters, from which the MMAD was calculated. The airborne density of the ferric oxide particles had previously been found to be 2.56 g/cm^3 [Spertell and Lippmann 1971].

Dual-Cast Exposure System

For each test run, two replicate casts were attached in series by their end branches, an arrangement similar to that used by Martin and Jacobi [1972]. A photograph of the exposure system is shown in Figure 1. In our study, distal branches were connected using lengths of latex tubing. The end branches of each cast had to be adapted for these tubes. To do this, short sections of latex tubing (~1 cm in length) were first attached to the airway ends of each cast using Dow Corning 3145 RTV. After the RTV cured, the tube sections were removed, resulting in circular-cross-section RTV airway ends on each hollow cast; these served as guides for extending the end airways for subsequent attachment of the latex connecting tubes between the two casts. The diameters of the latex tube sections were 0.32 cm (⅛ in.), 0.48 cm (³⁄₁₆ in.), 0.64 cm (¼ in.) or 0.79 cm (⁵⁄₁₆ in.); selection for each airway involved the smallest tube diameter that was larger than the midpoint diameter of that branch.

Before each test exposure, silicone oil (η = 0.5 Pa-sec) was coated on the

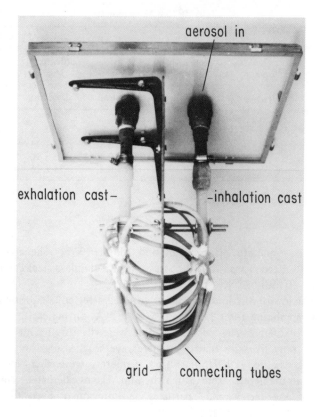

Figure 1.   Photograph of the dual replicate casts attached to the top of the "artificial thorax."

interior surface of the casts to simulate the wet TB surface. Particle deposition in the upper TB tree during inhalation is influenced by the larynx. To produce more realistic entry flow conditions, a hollow epoxy cast of a human larynx was attached to the trachea of the inhalation cast. A single vocal chord geometry was used in this study for all flowrates, avoiding the introduction of an additional variable affecting particle deposition; the glottic opening corresponded to that expected for a flowrate of 30 liter/min [Schlesinger et al. 1977].

The casts were then attached to fittings in the top of an acrylic "artificial thorax" [Schlesinger et al. 1977] via the larynx of the inhalation cast and the trachea of the exhalation cast. The latex connecting tubes were attached to matching airway ends using Dow Corning 738 RTV silicone rubber. This adhesive/sealant was chosen because it bonded more securely to the hollow cast material than to the latex tubing. While providing an airtight seal, the tubes could be removed easily from the airway ends following each exposure. To minimize movement of the tubes and any distortion of cast geometry, a 30.5- X 45.7-cm (12- X 18-in.) grid, consisting of a Masonite sheet with regularly spaced holes, was placed between the casts in the artificial thorax. The latex tubes were passed through specific locations in this grid so that their spatial orientation was consistent during each exposure (Figure 1). A gas leak detector was used before each test run to ensure airtight cast-tubing junctions.

The aerosol generator was attached to the artificial thorax opening at the inhalation cast. Aerosol was drawn through the casts using a vacuum pump that pulled at a constant flowrate of either 15 or 30 liter/min. Thus, in this manner, aerosol flowed from generations 0 through 4 in one cast (inhalation) and from 4 through 0 in the other (exhalation). A filter cartridge attached to the "thorax" opening at the exhalation cast collected particles that did not deposit in either cast.

Measurement of Deposited Activity

The method of intrabronchial deposition analysis has been described [Schlesinger et al. in press]. Basically, each hollow cast was marked off into specific counting regions before exposure. Length and bifurcations were individually delineated. (Each generation included the bifurcation(s) between it and the preceding generation [Weibel 1963]). The top of a bifurcation was taken to begin at the point where the distal portion of the preceding branch began to widen, and to end on a diagonal line ~1 mm below each carina; the trachea was marked into a number of 1-cm-long regions. After exposure, the cast was cut up along the marked regions, and each section was placed in a vial. These were counted using a 3-in.-diameter X 3-in. (7.6 X 7.6-cm) NaI (T1 activated) well counter.

Activity within the filter canister and latex connecting tubes were also measured so that an activity mass balance could be obtained. Activity within these was measured using a 3-in. diameter (7.5-cm) NaI scintillation detector housed in a low-background chamber. The activity of a point source of ^{99m}Tc was measured in a representative cast, connecting tubes and canister. The relative efficiency of each of the three counting geometries was then determined.

RESULTS AND DISCUSSION

A summary of test exposure conditions is presented in Table II. This table provides values for an impaction parameter, D^2Q, where D = particle

Table II. Summary of Test Exposures

Flowrate (liter/min)	Particle Size, MMAD (μm)	D^2Q[a]	Exhalation, ϵ_{exh}	Inhalation, ϵ_{inh}	R_ϵ[c]
			Deposition Efficiency, ϵ[b] (%)		
15	3.4 ± 1.16[d]	173	0.19	0.28	0.7
30	3.2 ± 1.14	307	0.16	0.20	0.8
30	4.3 ± 1.13	555	0.91	1.77	0.5
30	6.2 ± 1.16	1153	2.43	3.88	0.6

[a]This represents the product of flowrate and particle size.
[b]Efficiency is the percentage of aerosol entering the tracheobronchial tree cast which deposits in generations 0 through 3.
[c]$R_\epsilon = \epsilon_{exh}/\epsilon_{inh}$. The experimental error of R_ϵ is ±10%.
[d]Geometric standard deviation (σ_g).

MMAD (μ) and Q = flowrate (liter/min). Use of this parameter allows comparison of test runs performed at different particle sizes and flowrates. Previous work in this laboratory has shown that impaction is the dominant mechanism of deposition in the upper TB tree during inhalation for the particle sizes and flowrates used in this study [Lippmann 1977; Lippmann and Albert 1969; Lippmann et al. 1971; Schlesinger et al. 1977].

Table II also presents cast deposition efficiencies for inhalation and exhalation. Efficiency ϵ is defined as the percentage of aerosol entering the particular cast that deposits in generations 0 through 3. Although each cast extended through the fourth airway generation, deposition within this level is excluded from analysis, since the cast-tube junctions in this region may have produced artifacts in local flow patterns and, therefore, deposition patterns.

The ratio $R\epsilon$ of exhalation deposition efficiency ϵ_{exh} to inhalation deposition efficiency ϵ_{inh} for each test run is also presented in Table II. All ratios are <1, indicating that deposition within generations 0–3 during exhalation is less efficient than is that during inhalation. However, ϵ_{exh} is clearly a significant fraction of ϵ_{inh} in the casts for the particle size and flowrate conditions used in this study. The actual amount of exhalation deposition is, of course, dependent on the fraction of total aerosol that remains after inspiratory deposition occurs.

These results are consistent with those of Martin and Jacobi [1972] and Wang and Son [1977]. Martin and Jacobi studied the deposition of 0.2- to 0.4-μm (activity median diameter) radioactive aerosols in symmetrical airway models at constant flowrates ranging from 0.1 to 50 liter/min. They found that the probability of deposition during either inspiration or expiration was approximately equal for flowrates >1 liter/min through the casts. Using larger particles (1.14 and 2.05 μm, AED) and simulated respiratory flows, Wang and Son [1977] examined deposition during exhalation within a second-generation airway tube. They found that the average deposition velocities during exhalation were of the same order of magnitude as those during inhalation. Although flowrates were not reported, average air velocities in the region examined were 132 and 264 cm/sec (or ~17 and 34 liter/min total flow through the entire model if Weibel dimensions are assumed), which the authors reported corresponded to conditions at rest and with moderate exercise, respectively. Thus, based on these and the current study, exhalation and inhalation deposition efficiencies are of the same order of magnitude in the upper TB tree over a range of exposure conditions, i.e., particle sizes ranging from submicron to ~6 μm, and flowrates ranging from 1 to 30 liter/min.

Figure 2 shows deposition efficiencies for each generation during exhalation and inhalation. Peak efficiency during inhalation generally occurred in the third generation; this result is consistent with those obtained previously [Schlesinger et al. 1977]. During exhalation, the third generation exhibited peak efficiency in the 15- and 30-liter/min, 3.2-μm tests. For the two other, larger-size, particles at 30 liter/min, the first and second generations had relatively equal peak efficiency, with a rapid fall off in the next generation. Thus, peak efficiency during exhalation appears to be enhanced at the level of the first and second generations as flowrate and particle size increase, at least within the range of the parameters in this study.

Differences in the intraairway pattern of deposition between inhalation and exhalation were studied by examining the pattern of relative deposition in the region of the trachea (0 generation) and main bronchi (first generation). Figure 3 presents these data. In this figure, deposition in the designated regions is considered as 100%. Relative deposition was then defined as the percentage of the total that deposits in: (1) 1-cm segments of the trachea; (2) the tracheal bifurcation; (3) the right main bronchus; and (4) the left main bronchus.

Inhalation deposition is enhanced at the tracheal bifurcation and for several centimeters distal to the larynx, results consistent with those of earlier studies in this laboratory [Schlesinger and Lippmann 1976]. During exhalation, deposition is enhanced for several centimeters downstream (toward the larynx) of and at the tracheal bifurcation. Relative tracheal deposition generally decreases toward the larynx. As particle size increases, there is also a tendency for greater relative deposition in the main bronchi than in the bifurcation. This was not seen under any condition in the inhalation cast.

Previous studies [Schlesinger and Lippmann 1972; Schlesinger et al. 1977]

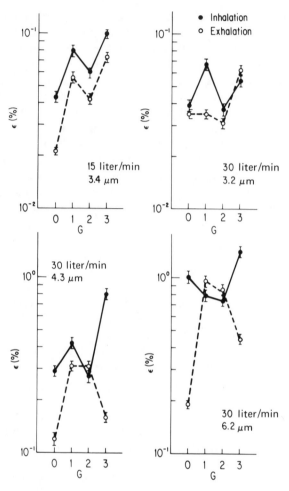

Figure 2.   Mean deposition efficiencies ϵ (%) for branching generations 0–3. (Note: differences in ordinate scales). Bars represent experimental errors.

have shown bifurcation regions to be sites of enhanced deposition during inhalation. Martin and Jacobi [1972] reported enhanced deposition of submicron particles at bifurcations during exhalation. Table III shows the fraction of deposition occurring within all bifurcation subregions through the third generation for the exhalation and inhalation casts. In general, the efficiency of deposition at bifurcations during inhalation increases with increasing impaction parameter; on the other hand, the efficiency of bifurcation deposition during exhalation decreases with increasing impaction

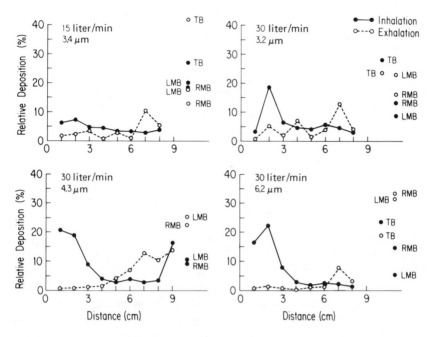

Figure 3.   Intratracheal deposition patterns. Distance is measured from the distal end of the cricoid cartilage (0 cm). Each point is plotted at the end of a 1-cm length along the trachea and represents the relative efficiency within that length. Relative deposition is obtained by designating total deposition in trachea and main bronchi as 100%, and taking percentages of this within the 1-cm tracheal regions, tracheal bifurcation (TB), right main bronchus (RMB) and left main bronchus (LMB).

parameter. Nevertheless, even during exhalation, deposition at bifurcations can account for >50% of the total deposition within the cast. It should be realized that length regions comprise more surface area than do bifurcation regions. Thus, even if there is <50% deposition at bifurcations, the surface density may be higher than that within the length sections. Therefore, surface density of deposition at bifurcations may be greater than that within lengths during exhalation as well as during inhalation.

The observation that for exhalation, the deposition percentage at bifurcations decreases as the impaction parameter increases suggests that impaction is not the mechanism accounting for deposition with these regions. However, in a converging air stream that changes direction at each proximal generation, the site of impaction would occur directly downstream of the region that would be measured experimentally as a bifurcation region; the site of enhanced deposition would then be counted as part of a length region.

Table III. Deposition at Bifurcation Regions

	Percent Total Deposition in Generations 0–3 That Occurs at Bifurcations	
D^2Q	Exhalation	Inhalation
173	55	38
307	56	46
555	46	58
1153	37	68

From Figure 3, it may be seen that there is enhanced deposition in regions downstream (toward the larynx) of the bifurcation during exhalation. Since bifurcation deposition decreases as D^2Q increases, then the relative deposition within length regions must increase as D^2Q increases. This relationship would not be due to deposition by sedimentation. Furthermore, sedimentation would result in deposition in the more distal half of the branch rather than the sections near the bifurcation; this is not observed in the trachea. Thus, if the pattern observed at and downstream of the tracheal bifurcation in the exhalation cast can be assumed to occur at other bifurcations, it appears that impaction does play a role in deposition during exhalation in the upper tracheobronchial tree, although the site of enhanced deposition may differ from that occurring in inhalation.

The results of this study indicate that bifurcations may be subjected to high local doses of inhaled particles during inhalation and exhalation. In addition, mathematical models that assume no deposition by impaction during exhalation may underestimate total collection efficiency.

CONCLUSIONS

This chapter has described a study of particle deposition during constant inhalation and exhalation in casts of the human upper TB tree. The efficiency of aerosol deposition during exhalation was 50–80% of that during inspiration. In addition, bifurcation regions may also be sites of selective deposition under exhalation; enhanced deposits occur downstream (toward the larynx) of carinal regions.

It should be emphasized that during an actual breathing cycle, flowrate varies from zero to a maximum. Deposition will be determined by the specific flowrates as well as by the length of the breath pause between inspiration and expiration.

ACKNOWLEDGMENTS

This study was supported by Grant No. ES 00881 from the National Institute of Environmental Sciences, and is part of a Center Program supported by NIEHS Center Grant No. ES00260. Mr. Concato was recipient of a NIOSH traineeship in occupational hygiene.

REFERENCES

Austin, E., J. Brock and E. Wissler (1979) "A Model for Deposition of Stable and Unstable Aerosols in the Human Respiratory Tract," *Am. Ind. Hygiene Assoc. J.* 40:1055.

Beeckmans, J. M. (1965) "The Deposition of Aerosols in the Respiratory Tract (1). Mathematical Analysis and Comparison with Experimental Data," *Can. J. Physiol. Pharmacol.* 43:157.

Findeisen, W. (1935) "Über das Absetzen Kleiner, in der Luft Suspendierten Teilchen in der Menschlichen Lunge bei der Atmung," *Arch. Ges. Physiol.* 236:367.

Hatch, T., and S. Choate (1929) "Statistical Description of the Size Properties of Nonuniform Particulate Substances," *J. Franklin Inst.* 207:369.

Landahl, H. D. (1950) "On the Removal of Airborne Droplets by the Human Respiratory Tract. I. The Lung," *Bull. Math. Biophys.* 12:43.

Lippmann, M. (1977) "Regional Deposition of Particles in the Human Respiratory Tract," in *Handbook of Physiology—Reactions to Environmental Agents* (Bethesda, MD: The American Physiological Society).

Lippmann, M., and R. E. Albert (1967) "A Compact Electric-Motor Driven Spinning Disc Aerosol Generator," *Am. Ind. Hygiene Assoc. J.* 28:501.

Lippmann, M., and R. E. Albert (1969) "The Effect of Particle Size on the Regional Deposition of Inhaled Aerosols in the Human Respiratory Tract," *Am. Ind. Hygiene Assoc. J.* 30:257.

Lippmann, M., R. E. Albert and H. T. Peterson (1971) "The Regional Deposition of Inhaled Aerosols in Man," in *Inhaled Particles III* W. H. Walton, Ed. (Old Woking, UK: Unwin Bros.).

Martin, D., and W. Jacobi (1972) "Diffusion Deposition of Small-Sized Particles in the Bronchial Tree," *Health Phys.* 23:23.

Schlesinger, R. B. (1980) "Particle Deposition in Model Systems of Human and Experimental Animal Airways," in *Generation of Aerosols and Facilities for Exposure Experiments* K. Willeke, Ed. (Ann Arbor, MI: Ann Arbor Science Publishers, Inc.).

Schlesinger, R. B., and M. Lippmann (1972) "Particle Deposition in Casts of the Human Tracheobronchial Tree," *Am. Ind. Hygiene Assoc. J.* 33:237.

Schlesinger, R. B., and M. Lippmann (1976) "Particle Deposition in the Trachea: *In vivo* and in Hollow Casts," *Thorax* 31:678.

Schlesinger, R. B., D. E. Bohning, T. L. Chan and M. Lippmann (1977) "Particle Deposition in a Hollow Cast of the Human Tracheobronchial Tree," *J. Aerosol Sci.* 8:429.

Schlesinger, R. B., J. L. Gurman and M. Lippmann (in press) "Particle Deposition Within Bronchial Airways: Comparisons Using Constant and Cyclic Inspiratory Flows," *Inhaled Particles V* (Oxford: Pergamon Press, Ltd.).

Schum, G. M., and H.-C. Yeh (1980) "Theoretical Evaluation of Aerosol Deposition in Anatomical Models of Mammalian Lung Airways," *Bull. Math. Biol.* 42:1.

Spertell, R. D., and M. Lippmann (1971) "Airborne Density of Ferric Oxide Aggregate Microspheres," *Am. Ind. Hygiene Assoc. J.* 32:734.

Wang, C.-S., and T. A. Son (1977) "Local Particle Deposition in Human Airway Models During Exhalation," paper presented at the American Industrial Hygiene Conference, Atlanta, GA, May 1977.

Weibel, E. R. (1963) *Morphometry of the Human Lung* (New York: Academic Press, Inc.).

Yu, C. P., and D. B. Taulbee (1977) "A Theory of Predicting Respiratory Tract Deposition of Inhaled Particles in Man," in *Inhaled Particles IV* W. H. Walton, Ed. (Oxford: Pergamon Press, Ltd.).

CHAPTER 15

MUCOCILIARY TRANSPORT AND PARTICLE CLEARANCE IN THE HUMAN TRACHEOBRONCHIAL TREE

C. P. Yu and J. P. Hu

Faculty of Engineering and Applied Sciences
State University of New York at Buffalo
Buffalo, New York

G. Leikauf, D. Spektor and M. Lippmann

Institute of Environmental Medicine
New York University Medical Center
New York, New York

ABSTRACT

The mucociliary clearance kinetics of inhaled particles have been measured by following the retention of a monodisperse Fe_2O_3 aerosol tagged with ^{99m}Tc. A compartmental model is proposed to describe the experimental result; from this model it is possible to determine mucociliary transport rate in each airway generation directly from the experimental retention curve. The calculated transport rate in the trachea is found to be within the range of experimentally observed values.

INTRODUCTION

Inhaled insoluble particles deposited in the lung are eliminated in time by several mechanisms. In the tracheobronchial (TB) or conducting airways, the most important mechanism is mucociliary clearance. The TB airways extend from the larynx to the terminal ciliated bronchioles, where mucociliary transport begins. Particles deposited on the surface of the mucus are transported

177

proximally as the mucus is propelled by the rhythmic beating of the cilia. The mucus eventually reaches the top of the trachea, at which point it is swallowed. Normally, mucociliary clearance is completed within 24 hours [Albert et al. 1967]; particle retention after 24 hours represents initial deposition beyond the ciliated airways and in the alveolar region of the lung. Clearance of particles in the alveolar region proceeds very slowly, and it may take several days or months [Morrow 1973].

One common method of evaluating mucociliary activity is to measure the retention of radiolabeled insoluble particles in the thorax in a sequence of times after inhalation [Albert et al. 1969]. A retention curve so obtained is found to be dependent on particle size and breathing condition, because of the change of particle deposition pattern. It has also been observed that there is reasonably good reproducibility of deposition and clearance in individual human subjects, but there is a substantial degree of intersubject variability [Albert et al. 1973]. Also, external agents such as drugs and inhaled irritants have been found to affect mucociliary transport rate significantly [Lippmann et al. 1977].

Because of the increasingly large airway surfaces with the airway generation number, it is believed that mucociliary transport rate decreases distally to prevent particle jam [Lourenco et al. 1971]. However, in vivo measurements of this rate have only been made in the trachea [Goodman et al. 1978; Yeates et al. 1975] and main bronchi [Foster et al. 1980]. In healthy subjects, there is a large scatter of observed tracheal velocity among individuals, and the results also depend on the experimental technique. The measurement values using invasive techniques are normally larger; they may reach as high as 20 mm/min. However, an average value of 5 mm/min was observed by non-invasive methods.

RETENTION MEASUREMENT

In this paper, clearance measurements are made for a subject using monodisperse Fe_2O_3 aerosol (aerodynamic diameter = 7.6 μm, σ_g = 1.1) tagged with ^{99m}Tc. The experimental apparatus and techniques for measuring TB clearance have been reported elsewhere [Albert et al. 1967,1969]. Before aerosol exposure, the subject performed a series of respiratory mechanical function tests. Then the subject inhaled orally the test aerosol for about one minute, regulating his own inspiratory flowrate (0.94 liter/min), tidal volume (1.5 liters) and respiratory rate (13.6 breaths/min).

Measurements of thoracic retention were made with two 5- X 2-in. NaI(Tl) scintillators positioned posterior and anterior to the upper right lung starting 5 min after inhalation and throughout the next 8 hr. An additional measurement was taken the following day. Figure 1 shows the clearance pattern of subject AN in the first 500 min.

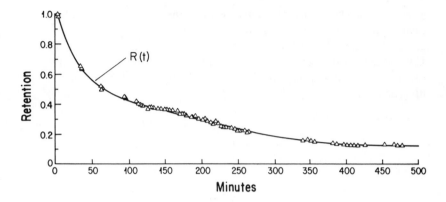

Figure 1.   Retention curve of subject AN.

Tracheal transport rates were determined with a probe consisting of a stack of six rectangular 1- X 4- X 2.75-cm collimated NaI(Tl) scintillation detectors suspended just above the carina and as close to the subject as possible. Sequential measurements were made while the subject sat in a low-background room with both detection systems in place. An average value of 5 mm/min tracheal velocity was observed.

MUCOCILIARY TRANSPORT MODEL

A compartmental model is proposed to describe mucociliary transport and particle clearance in the TB tree. This model suggests a procedure by which the mucociliary transport rate as a function of the airway depth can be determined from the retention curve R(t) in Figure 1 directly. We represent each airway generation by an escalator, moving with a velocity in the direction shown (Figure 2). For Weibel's [1963] lung model, there are 17 airway

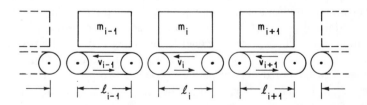

Figure 2.   Model of mucociliary transport.

generations in the tracheobronchial tree, and these are represented by 17 escalators with length ℓ_0 to ℓ_{16} and moving with velocity v_0 to v_{16}. Thus, generation 0 corresponds to the trachea and 1 to the main bronchi, and so on. Let the mass of the particles deposited in the ith generation be m_i and that in the nonciliated airways be M_P. Then, the mass fraction of the particles in the ith generation is

$$f_i = \frac{m_i}{M_{TB} + M_P} \tag{1}$$

where

$$M_{TB} = \sum_{i=0}^{16} m_i$$

For inhalation of a given aerosol source, m_i and M_P may be calculated using a deposition model developed earlier [Yu 1978; Yu and Diu 1982]. The mass balance of particles in the airways then yields:

$$R(T_n) = 1 - \sum_{i=0}^{n} f_i \quad n = 0 \ldots 16 \tag{2}$$

where $T_n = \sum_{i=0}^{n} t_i$, the time required to clear particles completely from the first n generations.

Equation 2 enables us to determine all values of t_i directly from the retention curve $R(t)$. The mucus velocity in the ith generation can then be found from:

$$v_i = \ell_i / t_i \quad i = 0 \ldots 16 \tag{3}$$

DEPOSITION PATTERN AND MUCOCILIARY TRANSPORT RATE

The deposition pattern along the respiratory airways depends on particle size, breathing condition and airway geometry. For the experimental condition described above, we have calculated the fractional deposition per airway generation using our deposition model by assuming that the airways of the subject follow the model of Weibel. Figure 3 gives the results of f_i for 7.6-μm particles. The fractional deposition clearly shows two peaks, corresponding to deposition by inertial impaction in the TB region and sedimentation in the pulmonary region. In the TB region, the maximum deposition occurs at generations 4 and 5.

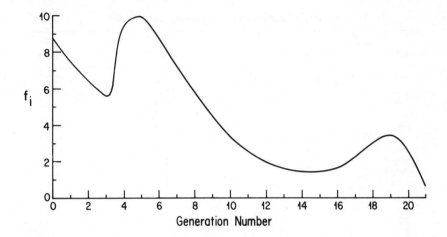

Figure 3.   Deposition fraction of 7.6-μm-diameter particles in the airways vs airway generations.

Following the procedure we outlined earlier, the values of mucociliary transport rate at each airway generation v_i are calculated from the retention curve R(t) shown in Figure 1 and deposition mass fraction f_i shown in Figure 3. The result is given in Table I and plotted in Figure 4. As expected, v_i tends to decrease with the increase of the generation number, but it has a nearly constant value between generations 7 and 14. The total time required to clear the TB region is found to be about 8 hr. This agrees with the experimental result. However, the tracheal transport rate obtained is considerably larger than the observed value of 5 mm/min for the subject, although it is still within the range of other reported data. This discrepancy may be due to the fact that the clearance measurements were made 5 min after the inhalation and that the dimensions of the subject's airways did not follow exactly Weibel's lung model. Figure 4 also shows mucociliary transport rates calculated by Lee et al. [1979] from the consideration of mass flow continuity of a mucous layer with constant layer thickness. Their results were based on:

$$v_i = 2^{-i}(d_0/d_i)v_0 \qquad (4)$$

where d_0 = trachea diameter
         d_i = ith generation airway diameter
         v_0 = observed tracheal transport rate, 5 mm/min

Equation 4 is independent of the shape of the retention curve R(t), which is obviously incorrect from physical grounds.

Table I. Calculated Mucociliary Transport Rate for Subject AN

Generation	ℓ_i (cm)	t_i (min)	v_i (mm/min)
0	12.0	9.2	13.0
1	4.76	5.3	8.98
2	1.90	5.5	3.45
3	0.76	5.0	1.52
4	1.27	13.5	0.94
5	1.07	20.3	0.53
6	0.90	31.7	0.28
7	0.76	50.8	0.15
8	0.64	45.7	0.14
9	0.54	38.0	0.14
10	0.46	30.0	0.15
11	0.39	28.0	0.14
12	0.33	25.0	0.13
13	0.27	22.5	0.12
14	0.23	19.5	0.11
15	0.20	45.0	0.04
16	0.165	81.0	0.02

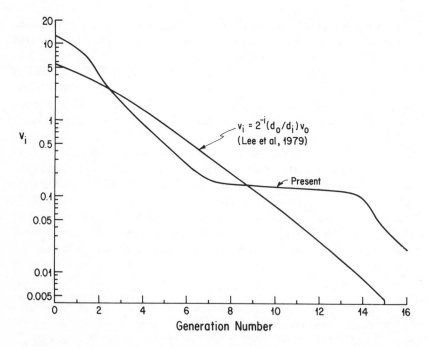

Figure 4.   Comparison of present mucociliary transport rate with the values from constant mucus layer model [Lee et al. 1979] by assuming tracheal velocity to be 5.5 mm/min.

It has been observed experimentally that, for identical particles, the retention curve differs markedly among individuals. Therefore, different mucociliary transport rates will be obtained for each subject from Equations 2 and 3. It has also been found that, even for the same subject mucociliary transport rates are still different if they are calculated from retention curves of different particle sizes. Since intrasubject variability is small, this inconsistency is caused by the assumption of Weibel's lung model. A study is in progress in which the current model is used as a tool to identify the airway dimensions such that the same mucociliary transport rate is found for an individual regardless what retention curves are used. The result of this study will be reported in a future publication.

ACKNOWLEDGMENTS

The work of Yu and Hu was supported by Grant No. ES-02565 from the National Institute of Environmental Health Sciences and Grant No. OH-00923 from the National Institute for Safety and Occupational Health. The work of Leikauf, Spektor and Lippmann was supported by Grant No. RP-1157 from the Electric Power Research Institute and by a center Grant ES-00260 from the National Institute of Environmental Health Sciences.

REFERENCES

Albert, R. E., M. Lippmann, J. Spiegelman, C. Strehlow, W. Briscoe, P. Wolfson and N. Nelson (1967) "The Clearance of Radioactive Particles from the Human Lung," in *Inhaled Particles and Vapour II*, C. N. Davies Ed. (Oxford: Pergamon Press, Ltd.), p. 361.

Albert, R. E., M. Lippmann and W. Briscoe (1969) "The Characteristics of Bronchial Clearance in Humans and the Effects of Cigarette Smoking," *Arch. Environ. Health* 18:738.

Albert, R. E., M. Lippmann, H. T. Peterson, Jr., J. Berger, K. Sanborn and D. Bohning (1973) "Bronchial Deposition and Clearance of Aerosols," *Arch. Intern. Med.* 131:115.

Foster, W. M., E. Langenback and E. H. Bergofsky (1980) "Measurements of Tracheal Bronchial Mucus Velocities in Man," *J. Appl. Physiol. Resp. Environ. Exercise Physiol.* 48:965.

Goodman, R. M., B. M. Yergin and J. F. Landa (1978) "Relationship of Smoking History and Pulmonary Function Tests to Tracheal Mucus Velocity in Nonsmokers, Young Smokers, Ex-smokers and Patients with Chronic Bronchitis," *Am. Rev. Resp. Dis.* 117:205.

Lee, P. S., T. R. Gerrity, F. J. Haas and R. V. Lourenco (1979) "A Model for Tracheobronchial Clearance of Inhaled Particles in Man and a Comparison with Data," *IEEE Trans. Biomed. Eng.* 26:624.

Lippmann, M., R. E. Albert and D. B. Yeates (1977) "Factors Affecting Tracheobronchial Mucociliary Transport," in *Inhaled Particles IV* W. H. Walton, Ed. (Oxford: Pergamon Press, Ltd.), p. 305.

Lourenco, R. V., M. F. Klimek and C. J. Borowski (1971) "Deposition and Clearance of 2 μ Particles in Tracheobronchial Tree of Normal Subjects— Smokers and Nonsmokers," *J. Clin. Invest.* 50:1411.

Morrow, P. E. (1973) "Alveolar Clearance of Aerosols," *Arch. Intern. Med.* 131:101.

Weibel, E. R. (1963) *Morphometry of Human Lung* (New York: Academic Press, Inc.).

Yeates, D. B., N. Aspin, H. Levison, M. T. Jones and A. C. Bryan (1975) 'Tracheal Transport Rates in Man," *J. Appl. Physiol.* 39:487.

Yu, C. P. (1978) "Exact Analysis of Aerosol Deposition During Steady Breathing," *Powder Technol.* 21:55.

Yu, C. P., and C. K. Diu (1982) "A Comparative Study of Aerosol Deposition in Different Lung Models," *Am. Ind. Hyg. Assoc. J.* 43:54.

CHAPTER 16

INHALABLE, INSPIRABLE AND TOTAL DUST

T. L. Ogden

Occupational Medicine and Hygiene Laboratories
Health and Safety Executive
London, United Kingdom

ABSTRACT

There has been considerable recent progress toward the definition of "total" dust for workplace sampling purposes. One line of approach has been to try to collect the biologically important fraction. It can be argued that, by this approach, it is sensible to collect the fraction of the total aerosol that enters the nose and mouth. This principle is incorporated in a report to be published by the International Standards Organisation (ISO) proposing that future standards should be based on the "inspirable" fraction. Biological importance is also the basis of the U.S. Environmental Protection Agency (EPA) "inhalable dust" definition, but EPA's proposal is not generally applicable in the workplace, because it takes insufficient account of head deposition and subsequent ingestion. An alternative approach to the biological one is to define a standard sampling method without prejudging what is thereby collected. A proposal to ISO based on this approach is that "total" dust should be defined as that collected by a sampling device into which air enters at a velocity of 1.1–3 m/sec, and in which the volume flowrate is 0.5–4 liter/min. Three recent computational studies enable us to estimate what particle-size range such a device would collect, provided that it is sharp-edged and operating in calm air. A particle of aerodynamic diameter d_a (cm) would be collected with >90% efficiency by a sharp-edged sampler diameter D (cm) and entry velocity V (cm/sec) in an external wind W (cm/sec) provided

$$d_a < 0.003 \, D^{0.2} \, V^{0.09}$$

and

$$W < 0.002 \, (D^2 V / d_a^4)^{1/3}$$

Thus, a sharp-edged sampler with this proposed ISO specification would efficiently collect particles up to about 40 μ aerodynamic diameter, but this would be limited to winds less than about 10 cm/sec. For blunt samplers, the diameter limit may be about half this. The theory for moving air is less well developed, and sampler shape would affect efficiency. We cannot, therefore, say what the ISO "total" dust proposals correspond to in moving air. However, experience gained from efficiency measurements on practical samplers should make it possible to make static and personal samplers that meet the "inspirable" specification, and the indications are that such a sampler would under most conditions collect more than one meeting the proposed "total" specification.

WHAT SHOULD BE MEASURED?

Until recently, reports of workplace measurements often implied that there was something called "total airborne dust" or "total suspended particulate" that could be measured simply by sucking air through a filter, without regard to the design of the filter holder. This is false on two counts. All particles fall relative to their surrounding air, and it is arbitrary whether one chooses as the upper limit of "airborne" particles, those with a terminal velocity of 1 mm/sec (6 μm aerodynamic diameter), 10 mm/sec (18 μm), 100 mm/sec (60 μm) or more. "Total dust" is only naturally defined if there is a particle size beyond which particles are so rare that they make negligible contribution to the aerosol mass (or whatever parameter is of interest). This condition is often not met in the workplace; thus, one must make some arbitrary definition. The second complication is that common samplers show aerodynamic entry effects that dictate what sizes they collect and make them dependent on any external air movement, so that measurement to any "total dust" definition is difficult.

In the last decade there have been two lines of approach to this problem for the purposes of standards (Figure 1). The first is to make a definition of "total dust" that reflects its biological effect. This approach has the outstanding advantage that measured concentrations are likely to correlate with disease: risks will be recognized earlier and control is more cost-effective. This approach is analogous to the use of respirable dust definitions

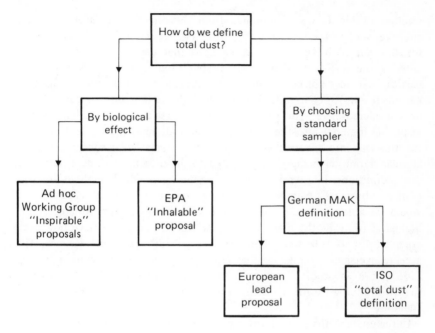

Figure 1.   Relationships between different conventions under discussion for defining 'total dust.'

for those substances that are most dangerous when they deposit deep in the lung. The second approach is to take a particular sampler design and define in terms of what that sampler collects. This, of course, cuts right across the problem of designing a sampler to match a specified characteristic, but it will only give good correlation with disease if the chosen design happens to match biological effect. This chapter considers the various proposals being discussed at present, and their relation to what samplers actually collect.

DEFINITIONS BASED ON BIOLOGICAL EFFECT

In 1979 a committee of the International Standards Organisation (ISO TC146) responsible for standards in the workplace and general environment set up a group to consider definitions of dust size ranges which might be used by TC146 in proposing measurement methods for particular substances. The group's proposals have been reported in full as a news item [Ad hoc Working Group 1981]. The proposals were based on possible biological effect, and the limit of what was considered relevant for the inspirable fraction (Table I) was determined from the directionally averaged entry characteristics of the human head [Armbruster et al. 1982]. The justification for this approach is that toxic materials that deposit immediately after entering the nose and mouth may be hazardous if they are swallowed; a particle that does not enter will clearly not present an inhalation or ingestion hazard. It may, of course, be absorbed through the skin, but this case will not be considered here. The proposals (Figure 2) went on to subdivide this inspirable material into the thoracic and extrathoracic (head) fractions, and the thoracic into the tracheobronchial (TB) and alveolar subfractions, the alveolar definition being closely related to the present workplace respirable dust definitions. ISO working parties proposing standard methods for particular substances, whether for the workplace or general environment, would be able to choose one of these subfractions as appropriate for the biological route of the material in question. The proposals have now been accepted by TC146 and will appear as an ISO Technical Report. The accepted version is similar to that reported by the Ad hoc Working Group [1981], but incorporates editorial and other minor changes. TC146 chose the 10-μ alternative for the division between the thoracic and extrathoracic fractions.

Unfortunately there is at present some confusion in terminology. "Inhalable fraction" was proposed at the 1975 Inhaled Particles IV conference, for material entering the nose and mouth [Ogden and Birkett 1977], and the term is still sometimes used this way (e.g., Armbruster et al. [1983]). Unfortunately, EPA adopted the same term with a different meaning, and the Ad hoc Working Group therefore chose "inspirable," to avoid confusion. This term is clearly preferred to avoid ambiguity.

The EPA definition of "inhalable" also is based on biological effect, but gives overwhelming emphasis to the material deposited below the larynx. Miller et al. [1979] concluded that only a few percent of 15-μm-aerodynamic-diameter particles deposited below the larynx. A proposed definition derived from this includes 50% of 15-μm particles in the inhalable fraction. This definition, therefore, has an implicit safety factor in relation to the lung

Table I. "Inspirable" Fraction, as Defined by the Ad Hoc Working Group

Particle Aerodynamic Diameter (μm)	0	4	12	20	50	100	185
Inspirable Fraction (%)	100	85.8	70.3	60.6	39.3	19.7	0

deposit, but seriously underrates material that may deposit in the nose and mouth and be swallowed. The EPA definition is therefore unsuitable for use in workplaces, if the material present may cause damage wherever it deposits, or may be absorbed in the gastrointestinal tract after swallowing.

DEFINITIONS BASED ON SAMPLER SPECIFICATION

An alternative approach is to specify the sampler used for measurement, without necessarily awaiting information on what sizes it collects (Figure 1). This procedure was adopted in the early 1970s by the German workplace standards (the MAK values), which defined total dust as that collected by a sampler with an entry velocity of 1.25 m/sec (±10%). This was in advance of the rest of the world at that time, but was incomplete because it is not only (or even chiefly) entry velocity that determines entry characteristics, and different sampler designs with the specified entry velocity could have different characteristics, especially if there were any external air movement. In practice, of course, only a small range of instruments were used [Schutz and Coenen 1974]. This definition has been widely adopted for personal sampling in continental Europe, for example by operating a field monitor cassette with a 4-mm entry orifice at the appropriate flowrate, about 0.9 liter/min. The entry velocity of 1.25 m/sec is at present being considered as part of a European communities standard for lead in the workplace, with the additional specification of an entry orifice diameter of 4-10 mm.

A similar path is being followed by another working group of ISO TC146. The current proposal of this group is to define "total dust" as that collected by a sampler with an entry velocity between 1.1 and 3 m/sec, at flowrates between 0.5 and 4 liter/min, corresponding to a diameter range of circular orifices of 2-9 mm. This is not fully consistent with the Ad hoc Working Group approach, because what will be collected at these entry velocities does not necessarily correspond with any of the Ad hoc Working Group subfractions. However, the Ad hoc Working Group did not consider instrumentational realization of its proposals (Armbruster et al. [1983] discuss some possibilities), and its definitions should be seen as a target for future standards and instrument design. The entry velocity approach gives immediate operational guidance, but it is to be hoped that for the sake of consistency it will be superseded in the medium-term by instrumentation based on the Ad hoc Working Group proposals on biological effects. The rest of this chapter will consider what design variables of samplers determine their entry efficiencies, and therefore, how results using these different definitions are likely to be related.

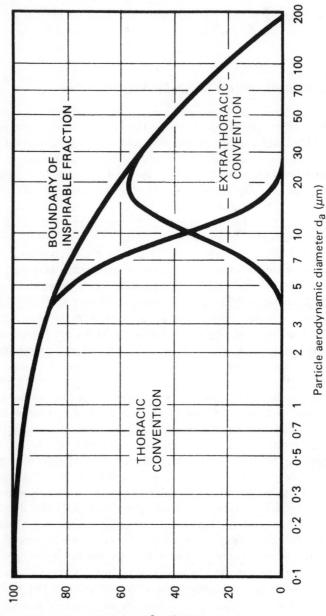

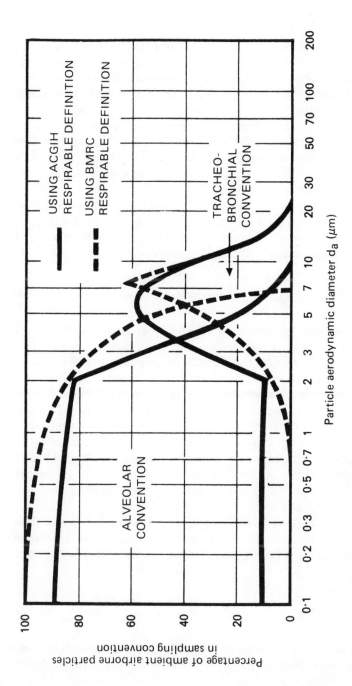

Figure 2. Conventions proposed to International Standards Organisation Technical Committee 146 by the Ad hoc Working Group for use in sampling airborne particles. An alternative alveolar convention, similar in shape to the ACGIH but with all the diameter values reduced by 29%, is included in the proposals for cases where the target population is the sick or infirm rather than healthy adults.

SAMPLER EFFICIENCIES IN CALM AIR

It was shown by Davies [1968] that the entry velocity V of a small sharp-edged sampler should exceed 25 times the sedimentation velocity v of the largest particle it is intended to measure. This would lead to an upward facing sharp-edged sampler oversampling by 4% and a downward facing sampler undersampling by the same amount relative to the true concentration. This sedimentation criterion is a minimum requirement in the design of any sampler entry for calm air. The entry velocity of 1.25 m/sec, mentioned in the previous section, would, according to this criterion, be good for aerodynamic diameters up to about 40 μm. At larger sizes, sampler performance would depend increasingly on orientation.

An additional limit occurs because the sedimentation velocity and inertia of particles prevent them from perfectly following the air flowlines: this limit can be expressed as a criterion for minimum entry orifice diameter D. Davies [1968] proposed:

$$D > 62.5 \; Vv/g \tag{1}$$

However, this work has been superseded by three recent computations of the critical trajectories of particles into thin tubes. Before comparing these, it should be noted that all three studies assume that Stokes' law can be used to calculate particle drag; however, such an assumption will generally lead to an overestimate of sampling error. For example, the true drag on a sphere of 25-μm diameter with a relative velocity of 2 m/sec is 30% larger than that calculated from Stokes' law [Davies 1945]. There is no difficulty in computing critical trajectories using the true drag values, but the Stokes' law assumption is, of course, very convenient because it allows the results to be expressed in terms of simple and convenient criteria for the sampling parameters. However, the approximations involved in deriving the criteria means that they must be applied with caution.

The results of the three recent computational studies are presented differently in the original papers, but comparable expressions can be derived from them. Agarwal and Liu [1980] conclude that, for entry efficiency to exceed 90%,

$$D > 20v^2/g \tag{2}$$

Yoshida et al. [1978] presented their results as contours of equal efficiency plotted on axes of Stokes' number and relative terminal velocity. The 90%

contour can be fitted by a curve, giving the requirement for efficiency to exceed 90% as

$$D > 99v^{2.43}/gV^{0.43} \qquad (3)$$

For comparison with Equation 2, a less exact fit to the results of Yoshida et al. can be obtained, constrained to make D independent of V:

$$D > 36v^2/g \qquad (4)$$

The third author's results [ter Kuile 1979], can be expressed as

$$D > 78v^{2.5}/gV^{0.5} \qquad (5)$$

All of these results can, of course, be roughly related to particle diameter d using the Stokes' law equation

$$v = \sigma d^2 g/18\eta \qquad (6)$$

where η = dynamic viscosity of air
         σ = particle density

The differences in these expressions probably reflect the different ways in which the criteria are derived from the calculated trajectories, rather than differences in the trajectories themselves. The minimum diameters predicted by these approximate criteria are, however, substantially different, with Agarwal and Liu's formulas giving usually the largest entry diameter (i.e. most stringent requirement) for small particles, and Equation 3 (based on Yoshida) predicting the highest minimum diameter for larger particles and lower flowrates. The ter Kuile formula is always the least strict, i.e., it allows smaller tube diameters. The differences are illustrated in Table II for an entry velocity V of 300 cm/sec.

In practice, it might be unsafe to accept the most liberal of these criteria, that of ter Kuile. Expression 2 above is the easiest for quick calculation, but Equation 3 is probably a more accurate representation of much the same results. Table III, therefore, gives minimum entry diameters calculated from the work of Yoshida et al. using Expression 3. It illustrates that the minimum required entry diameter decreases slowly as intake velocity increases, but increases very rapidly with d_a, approximately as d_a^5. We may

Table II. Minimum Orifice Entry Diameters (cm) of Sharp-Edged Samplers
for an Entry Velocity of 300 cm/sec, Predicted by Three Criteria

	Particle Aerodynamic Diameter (μm)			
	20	30	40	60
A[a]	0.030	0.15	0.48	2.4
Y[b]	0.014	0.10	0.40	2.9
K[c]	0.007	0.06	0.24	1.8

[a]A = Agarwal and Liu (Expression 2).
[b]Y = Yoshida et al. (Expression 3).
[c]K = Ter Kuile (Expression 5).

Table III. Minimum Orifice Entry Diameters (cm) for Sharp-Edged
Sampler Entry Efficiencies to Exceed 90% in Calm Air

Sampler Entry Velocity V (cm/sec)	Particle Aerodynamic Diameter d_a (μm)				
	20	30	40	60	100
30	0.037	a	a	a	a
100	0.022	0.16	a	a	a
125	0.020	0.15	0.59	a	a
300	0.014	0.10	0.40	2.9	a
1000	0.008	0.06	0.24	1.7	21

[a]Davies' sedimentation criterion not met (V < 25v).

conclude that, with sharp-edged samplers in calm air, it is very difficult to collect 100-μm particles efficiently, but difficult not to collect 20-μm particles.

It must be remembered, however, that Table III relates to sharp-edged samplers in calm air. Workplace air is, of course, never truly still, and slight air movement will reduce efficiency, even for a sampler facing the wind. It is possible to assign an upper limit of windspeed W for calm air, using the equation of Levin [1957] in the form given by Belyaev and Levin [1974]. This equation appears to be approximately correct for sharp-edged samplers facing the wind [Gibson and Ogden 1977] and to underestimate the error for samplers side-on to the wind [Davies and Subari 1978]. If Stokes' law is assumed, then the windspeed W beyond which intake efficiency falls below E is given by:

$$W = 0.01(1 - E)^{2/3}(D^2 V/d_a^4)^{1/3} \qquad (7)$$

Figure 3 illustrates the limiting windspeeds for efficiency to exceed 90% with intake velocities of 1.25 and 2.5 m/sec. As already noted, it is not advisable to try to collect particles >40 μm with an intake velocity of 1.25 m/sec, because Davies' sedimentation criterion is not met; efficiency will then depend appreciably on sampler orientation. It is convenient to take this limiting windspeed for E = 0.9 as the practical limit of calm air conditions for the values of D, V and d_a used, and to assume that under these conditions one can apply the calm air expressions (Equations 2 to 5). This is, of course, only a crude approximation because of the assumption of Stokes' law, and because the full approach would require calculation of critical trajectories in moving air. Until this is done, the best procedure in testing sampling conditions is probably to check that calm air conditions apply, i.e., that W is below the value given by Equation 7 for E = 0.9, and then to check that Davies' sedimentation criterion (V > 25v) and the inertial criterion of Expression 3 (Table III) or the simpler one of Expression 2 are met.

The other limitation to the application of Expressions 2 to 5 is sampler bluntness. Trajectories have been computed for a few sampler designs, but general conditions have not been given. There are also few actual measurements of efficiency below 95% in calm air, let alone the 90% required to test the sharp-edged sampler Equations 2 to 5. Table IV summarizes results available for 90% efficiency, together with diameters predicted by Yoshida et al. at this efficiency level.

The experimental values in Table IV are all rough, but are consistently below the predictions. It seems likely that this is due to sampler bluntness.

Table IV. Measurements of Maximum Particle Size for 90%
Efficiency for Practical Blunt Samplers, Compared with
Predictions Derived from Yoshida et al. [1978]
for Sharp-Edged Samplers

Minimum Orifice Dimension D (cm)	Entry Velocity V (cm/sec)	Aerodynamic Diameter for 90% Efficiency (μm)		Reference
		Predicted	Measured	
0.025	40	18	9	Ogden et al. [1978]
0.04	6000	32	10	Breslin and Stein [1975]
0.12	4400	40	4	Breslin and Stein [1975]
0.25	10	26	20	Ogden et al. [1978]
0.25	20	27	18	Ogden and Birkett [1978]

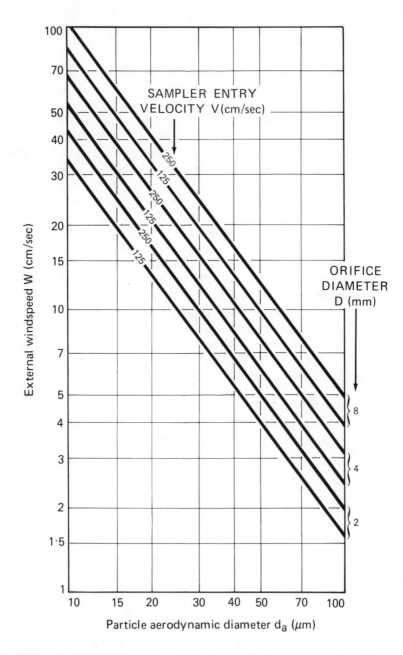

Figure 3.   Proposed limits for application of "calm air" theories. The theories apply for combinations of external windspeed W and particle aerodynamic diameter d_a, which lie to the left of the lines defined by the sampler. Doubling the sampler flowrate increases the limiting windspeed by 26%.

The designs tested all approximate to holes in relatively large, flat areas, which might give enhanced possibility of impaction before collection when compared with the sharp-edged samplers of the theory. There have been other experiments that have not extended to large enough particle diameters to give efficiencies <90%, but the results are consistent with this picture. These results are summarized in Table V.

To sum up, calm air criteria are likely to be applicable if the wind speed W (cm/sec) is:

$$W < 0.002(D^2 V/d_a^4)^{1/3} \tag{8}$$

where   D = sampler minimum entry dimension (cm)
        V = mean intake velocity (cm/sec)
        d_a = particle aerodynamic diameter (cm)

This equation is obtained by putting E = 0.9 in Equation 7. If this requirement is met, then, using Equations 3 and 6, the maximum aerodynamic diameter for efficiencies to exceed 90% at 20°C is given roughly by:

$$d_a < 0.003\ D^{0.2}\ V^{0.09} \tag{9}$$

For blunt samplers, the limiting value of d_a may be less than half that given by Equation 9. If a definition is required for the operating limits of a particular sampler, then, of course, the limiting d_a must be obtained first, from Expression 9, and this value used to test the windspeed in Expression 8.

Table V. Sampler Studies That Gave Efficiencies >90% at All Measured Sizes, Compared with Predictions Derived from Yoshida et al. [1978] for Sharp-Edged Samplers

Minimum Orifice Dimension D (cm)	Entry Velocity V (cm/sec)	Aerodynamic Diameter		Reference
		Predicted for 90% Efficiency (μm)	Maximum Tested (μm)	
0.2	840	38	8	Pickett and Sansone [1973]
0.3	4700	50	40	Gibson and Ogden [1977]
0.4	400	42	20	Breslin and Stein [1975]
0.5	80	37	20	Ogden and Birkett [1978]
0.65	550	48	30	May and Druett [1953]

SAMPLERS IN MOVING AIR

It is well known that a sharp-edged sampler facing the wind has an efficiency described by

$$E = W/V + K(1 - W/V) \qquad (10)$$

where K is a function of Stokes' number and W/V. There is an extensive literature on the exact form of K, but the relationship between E and W/V has the general shape shown in Figure 4a. For samplers not facing the wind there is much less information [Durham and Lundgren 1980; Willeke and Tufto 1983].

Figure 4b illustrates the corresponding relationship between E and W/V for a blunt sampler with a (roughly) central orifice facing the wind. A typical set of results is given by May and Druett [1953], but many other studies of blunt samplers have a similar pattern. It will be seen that, unlike the sharp-edged case, there is no windspeed for which efficiency is unity for all particles, and it follows that efficiency cannot be fully characterized by Stokes' number and the velocities. Moreover, in the crossover region (A in Figure 4b), efficiency can increase with particle diameter at small diameters and decrease at large diameters [Ogden and Birkett 1977], or the reverse [May and Druett 1953]. These features are qualitatively explicable from the flow-pattern in front of the blunt sampler, which is typically a region of divergence followed by a region of convergence. Vincent and Mark [in press] have adapted Equation 10 to take account of this double pattern, introducing the sampler bluntness B. Their new expression can account for the features of the efficiency curves, but theory is not yet sufficiently developed for the efficiency of blunt samplers to be calculated from their dimensions, W and V, as it can with sharp-edged samplers. In any case, there is some evidence that, if the orifice is small compared with the overall diameter of the sampler, intake flowrate, rather than velocity, is the important parameter [Ogden and Birkett 1977]. A commonly observed feature of blunt samplers is that as W/V increases from zero, efficiency changes are initially rapid, but toward the right side of Figure 4b (typically $W > 2$ m/sec), change with W/V is slow.

As a blunt sampler is turned away from facing into the wind, efficiency decreases until the orifice is side-on to the wind, and as angle is further increased, efficiency may decrease further, stay about the same, or increase, depending on factors yet to be clarified. In the side-on position, increasing W or d_a usually will decrease efficiency. If an orifice is markedly off-center, the air is likely to be deflected by the bulk of the instrument, so that the flow is past the orifice and the efficiency follows the side-on pattern even when the instrument as a whole is facing the wind [Ogden et al. 1978]. An

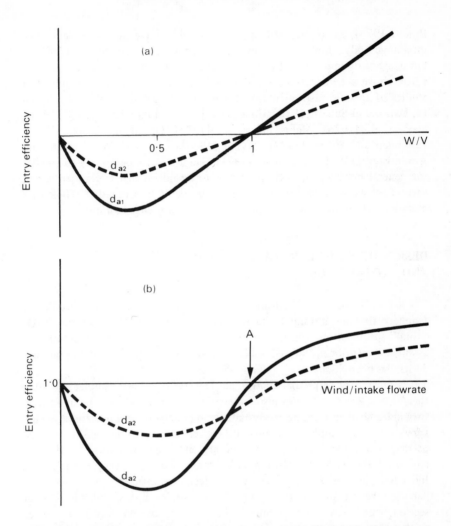

Figure 4.   General forms of the relations between entry efficiency, windspeed and entry velocity or flowrate for (a) a sharp-edged sampler facing the wind, (b) a blunt sampler with central orifice facing the wind. d_{a1} and d_{a2} are two particle aerodynamic diameters; $d_{a1} > d_{a2}$.

important feature is that, away from the facing-wind position, sampler efficiency is much less dependent on W/V than it is in the facing position.

An important question for the workplace is the dependence of a personal sampler attached to a body on external air movement, due either to wind or the movement of the wearer. Information on this is very limited [Wood and

Birkett 1979], but a sampler may then behave like a small off-center orifice on a large body, showing a decline in efficiency with increasing particle size, but being fairly independent of the wearer's orientation to the wind.

The above generalizations are derived from the large number of published studies of specific instruments, including, apart from those already referred to, Davies and Subari [1978], Hirst [1952], Liu and Pui [1981], May [1956], May et al. [1976], McFarland et al. [1977], Ogden and Wood [1975], Pattenden and Wiffen [1977], Raynor [1970], Sehmel [1970], Vincent and Armbruster [1981] and Zenker [1971]. These results permit us to predict the general behavior of a sampler; it is to be hoped that experimental and theoretical work currently in progress will enable us to predict efficiencies quantitatively in the way that can be done for sharp-edged samplers.

DESIGN OF SAMPLERS FOR INSPIRABLE AND "TOTAL" DUST

Is it possible to use the information in the previous two sections to design samplers for the inspirable fraction, or to say what the ISO and MAK "total" specifications will in fact measure? (The EPA inhalable fraction has, as already mentioned, limited application in the workplace; suitable sampler design has been discussed by Liu and Pui [1981].)

It is fairly easy to predict the general features of a static sampler matching the Ad hoc Working Group inspirable curve. The sampler must be insensitive to wind and, therefore, probably radially symmetrical in a horizontal plane. Upward-facing samplers are obviously excluded to avoid fall-in of large particles; downward-facing samplers are also unsatisfactory because in any wind their efficiency often falls rapidly with increasing particle size. Independence of wind direction and slow decline of efficiency with increasing particle size are probably best met by a sampler with circular horizontal section, whose entry is a circumferential slit or close series of holes, offset so that flow past the entry is always side-on to the entry even on the windward side of the sampler. These features are seen in the Orb sampler [Ogden and Birkett 1978], which has been shown to follow the inspirable curve over a limited part of its range. The Orb sampler is made fairly independent of windspeed by having an obstruction on the local downwind side of the entry holes, so that the flow near the entry is close to stagnant, and this may be a generally useful feature [May 1967]. Vincent and Gibson [1981] have shown that the entry efficiencies of the Orb depend on the degree of adhesion of the particles near the entry, and future designs should clearly seek to remove this dependence. The fit of other designs to the inspirable curve is discussed by Armbruster et al. [1983].

As already mentioned, there is little information on the aerodynamics of personal samplers mounted on the lapel, but there is reasonable hope that they can be made to be independent of windspeed and to show the slow decline with particle size of the inspirable curve. It can be argued that it is more rational to make a personal sampler that imitates the detailed directional dependence of the wearer's entry efficiency, rather than seeking to maintain a directionally averaged curve independently of wind direction. This may, however, be more difficult, and there is in any case better agreement on the directionally averaged curve than on the values at particular angles [Vincent and Armbruster 1981]. It is clear from the foregoing that it is no easier to make a sampler to measure total dust independent of wind speed and direction than it is to make one following the inspirable curve.

Turning to the ISO "total dust" proposal discussed above, the currently proposed ranges of flowrate (0.5–4 liter/min) and entry velocity (1.1–3 m/sec) would allow a variation in the limiting aerodynamic diameter (from Expression 9) between 36 and 45 μm. Despite the considerable range of choice of these variables, therefore, efficiencies of sharp-edged samplers meeting the specifications are likely to be similar in calm air. Blunt samplers, however, will have efficiencies that start to fall at very roughly half these particle diameters. Moreover, limiting windspeed for these calm air conditions to apply is found from Expression 8 to be 8–12 cm/sec. Therefore, these calm air efficiencies are unlikely to apply very often in the workplace, where air will hardly ever be this still. However, Expression 8 shows that if the particles present are restricted to small sizes, then a greater range of velocity can be permitted. For 10-μm particles, sharp-edged samplers meeting the proposed ISO specification will give efficiencies greater than 90% for windspeeds below 45 cm/sec for the smallest permitted entry diameter and below 85 cm/sec for the largest permitted entry diameter.

At higher windspeeds, theory does not permit us to be this specific, but for the important personal sampling case the work of Wood and Birkett [1979] gives some information. One of the samplers they tested (the United Kingdom Atomic Energy Authority head) has a 4-mm orifice (roughly the same as the 37-mm air monitor) and was used at 2 liter/min (i.e., an entry velocity of 2.7 m/sec). It showed an efficiency falling to about 50% at 15 μm and 10% at 35 μm when the wearer faced a wind of 1 m/sec. As already discussed, other orientations and higher windspeeds are likely to give efficiencies roughly the same or perhaps somewhat lower. It is, therefore, likely that in modest windspeed (or if the wearer is walking) the ISO "total dust" specification will undersample larger particles relative to the inspirable definition— the opposite of what might be thought at first sight.

CONCLUSIONS

There has been progress in the last two or three years toward agreed definitions for "total" dust, based either on the likely biological effect of the aerosol or on the choice of a standard sampler. Available evidence enables us to make rough predictions of what these standard samplers will collect in calm air, to define the limits of applicability of these predictions, and to estimate the efficiencies of sharp-edged samplers in moving air. We cannot at present estimate quantitatively the efficiency of blunt samplers in moving air, but we can make qualitative statements about which variables are important, and rapid progress is being made. We should, therefore, soon be able to collect the inspirable fraction with personal samplers, and also be able to say what would be the result of using samplers that meet other specifications. Sampling the inspirable fraction would be the best long-term choice, as this is likely to give the closest relationship between environmental monitoring and biological monitoring or disease.

REFERENCES

Ad hoc Working Group to Technical Committee 146—Air Quality, International Standards Organisation (1981) "Size Definitions for Particle Sampling," *Am. Ind. Hygiene Assoc. J.* 42:A64-A68.

Agarwal, J. K., and B. Y. H. Liu (1980) "A Criterion for Accurate Aerosol Sampling in Calm Air," *Am. Ind. Hygiene Assoc. J.* 41:191-197.

Armbruster, L., H. Breuer, J. H. Vincent and D. Mark (1983) "Definition and Measurement of Inhalable Dust," Chapter 17, this volume.

Belyaev, S. P., and L. M. Levin (1974) "Techniques for Collection of Representative Aerosol Samples," *J. Aerosol Sci.* 5:325-338.

Breslin, J. A., and R. L. Stein (1975) "Efficiency of Dust Sampling Inlets in Air," *Am. Ind. Hygiene Assoc. J.* 36:576-583.

Davies, C. N. (1945) "Definitive Equations for the Fluid Resistance of Spheres," *Proc. Phys. Soc.* 57:259-270.

Davies, C. N. (1968) "The Entry of Aerosols into Sampling Tubes and Heads," *Brit. J. Appl. Phys.* 2s 1:921-932.

Davies, C. N., and M. Subari (1978) "Inertia Effects in Sampling Aerosols," paper presented at the Symposium on Advances in Particle Sampling, Asheville, North Carolina, May 1978.

Durham, M. D., and D. A. Lundgren (1980) "Evaluation of Aerosol Aspiration Efficiency as a Function of Stokes Number, Velocity Ratio and Nozzle Angle," *J. Aerosol Sci.* 11:179-188.

Gibson, H., and T. L. Ogden (1977) "Entry Efficiencies for Sharp-Edged Samplers in Calm Air," *J. Aerosol Sci.* 8:361-365.

Hirst, J. M. (1952) "An Automated Volumetric Spore Trap," *Ann. Appl. Biol.* 39:257-265.

Levin, L. M. (1957) "The Intake of Aerosol Samples," *Izv. Akad. Nauk SSSR Ser. Geofiz.* 7:914-925.

Liu, B. Y. H., and D. Y. H. Pui (1981) "Aerosol Sampling Inlets and Inhalable Particles," *Atmos. Environ.* 15:589-600.

May, K. R. (1956) "A Cascade Impactor with Moving Slides," *Am. Med. Assoc. Arch. Environ. Health* 13:481-488.

May, K. R. (1967) "Physical Aspects of Sampling Airborne Microbes," *Symp. Soc. Gen. Microbiol.* 17:60-80.

May, K. R. and H. A. Druett (1953) "The Pre-impinger: A Selective Aerosol Sampler," *Brit. J. Ind. Med.* 10:142-151.

May, K. R., N. P. Pomeroy and S. Hibbs (1976) "Sampling Techniques for Large Windborne Particles," *J. Aerosol Sci.* 7:53-62.

McFarland, A. R., J. B. Wedding and J. E. Cermak (1977) "Wind Tunnel Evaluation of a Modified Andersen Impactor and an All Weather Sampler Inlet," *Atmos. Environ.* 11:535-539.

Miller, F. J., D. E. Gardner, J. A. Graham, R. E. Lee, W. E. Wilson and J. D. Bachmann (1979) "Size Considerations for Establishing a Standard for Inhalable Particles," *J. Air Poll. Control Assoc.* 29:610-615.

Ogden, T. L., and J. L. Birkett (1977) "The Human Head as a Dust Sampler," in *Inhaled Particles IV* W. H. Walton, Ed. (Oxford: Pergamon Press, Ltd.), pp. 93-105.

Ogden, T. L., and J. L. Birkett (1978) "An Inhalable-Dust Sampler, for Measuring the Hazard from Total Airborne Particulate," *Ann. Occup. Hygiene* 21:41-50.

Ogden, T. L., and J. D. Wood (1975) "Effects of Wind on the Dust and Benzene-Soluble Matter Captured by a Small Sampler," *Ann. Occup. Hygiene* 17:187-195.

Ogden, T. L., J. L. Birkett and H. Gibson (1978) "Large-Particle Entry Efficiencies of the MRE 113A Gravimetric Dust Sampler," *Ann. Occup. Hygiene* 21:251-263.

Pattenden, N. J., and R. D. Wiffen (1977) "The Particle Size Dependence of the Collection Efficiency of an Environmental Aerosol Sampler," *Atmos. Environ.* 11:677-681.

Pickett, W. E., and E. B. Sansone (1973) "The Effect of Varying Inlet Geometry on the Collection Characteristics of a 10 mm Nylon Cyclone," *Am. Ind. Hyg. Assoc. J.* 34:421-428.

Raynor, G. S. (1970) "Variation in Entrance Efficiency of a Filter Sampler with Air Speed, Flow Rate, Angle and Particle Size," *Am. Ind. Hygiene Assoc. J.* 31:294-304.

Schutz, A., and W. Coenen (1974) "Feinstaub: Definition—Messverfahren," *Staub* 34:323-326.

Sehmel, G. A. (1970) "Particle Sampling Bias Introduced by Anisokinetic Sampling and Deposition Within the Sampling Line," *Am. Ind. Hygiene Assoc. J.* 31:758-771.

ter Kuile, W. M. (1979) "Dust Sampling Criteria," *J. Aerosol Sci.* 10:241-242.

Vincent, J. H., and L. Armbruster (1981) "On the Quantitative Definition of the Inhalability of Airborne Dust," *Ann. Occup. Hygiene* 24:245-248.

Vincent, J. H., and H. Gibson (1981) "Sampling Errors in Blunt Dust Samplers Arising from External Wall Loss Effects," *Atmos. Environ.* 15:703-712.

Vincent, J. H., and D. Mark (in press) "Applications of Blunt Sampler Theory to the Definition and Measurement of Inhalable Dust," in *Inhaled Particles V*, (Oxford: Pergamon Press, Ltd.).

Willeke, K., and P. Å. Tufto (1983) "Sampling Efficiency Determination of Aerosol Sampling Inlets," Chapter 25, this volume.

Wood, J. D., and J. L. Birkett (1979) "External Flow Effects on Personal Sampling," *Ann. Occup. Hygiene* 22:299-310.

Yoshida, H., M. Uragami, H. Masuda and K. Iinoya (1978) "Particle Sampling Efficiency in Still Air," *Kagaku Kogaku Robunshu* 4: 123-128.

Zenker, P. (1971) "Untersuchungen zur Frage der Nichtgeschwindigkeitsgleichen Teilstromentnahme bei der Staubgehaltsbestimmung in Stromenden Gasen," *Staub* 31:252-256.

CHAPTER 17

DEFINITION AND MEASUREMENT OF INHALABLE DUST

L. Armbruster and H. Breuer

Steinkohlenbergbauverein
Essen, West Germany

J. H. Vincent and D. Mark

Physics Branch
Institute of Occupational Medicine
Edinburgh, United Kingdom

ABSTRACT

This chapter describes work that has been carried out in recent years in Europe to develop criteria for the measurement of inhalable dust in workplaces—that is, the biologically relevant fraction of airborne dust that enters the body during the act of breathing. Quantitative definition of the inhalable dust fraction (or inhalability) was obtained from experiments with models (life-size tailor's dummies) in wind tunnels. Simple blunt sampler theory allowed interpretation of these results and provided the basis for a new generation of dust sampling inlets for collecting inhalable dust. A number of sampling inlets already in existence meet the main geometrical criteria and, when tested, were found to have entry characteristics that matched inhalability to a greater or lesser extent. Finally, the role of inhalable dust in good dust sampling practice in occupational hygiene is discussed.

INTRODUCTION

In most instances, the only particles of airborne dust that could be harmful to a human subject are those that enter the body through the nose and/or mouth during the act of breathing and subsequently deposit somewhere in the respiratory tract. The part of airborne dust that enters is referred to here as "inhalable dust," and the fraction as a function of particle aerodynamic diameter as "inhalability." This terminology has been used in Europe for a number of years, and is not to be confused with the U.S. Environmental Protection Agency (EPA) definition of "inhalable particulate" [Miller et al. 1979], which refers only to those particles that penetrate into the lower respiratory tract.

In occupational medicine, it is necessary to understand how a given regional deposition subfraction—nasopharyngeal, tracheobronchial (TB) or respirable—of the inhaled fraction relates to a given disease. Since dust sampling in practice is usually carried out in the workplace for the express purpose of assessing potential risks to the health of the workforce, good sampling practice ought to reflect recognition of the physical picture embodied in this last statement. This chapter describes recent work carried out in West Germany and in the United Kingdom that sets out to provide the basis for sampling inhalable dust and to examine the performances of a number of sampling heads that appear to be potentially suitable for practical use.

QUANTITATIVE DEFINITION OF INHALABLE DUST

Experiments for quantifying inhalable dust were carried out by Ogden and Birkett [1977] in a wind tunnel of 0.5 m^2 cross section, using a full-size model human head with nose and mouth openings of typical shape and dimensions. Filter holders were located behind the mouth and nose openings in such a way as to allow sampling of the inhaled aerosol without losses. The investigation covered a particle aerodynamic diameter range up to 30 μm, wind speeds up to 2.75 m/sec, and orientations of the head with respect to the wind up to 180°. More recently, Armbruster and Breuer [1982], in Essen, West Germany, carried out investigations in a similar test arrangement (Figure 1) over more comprehensive ranges of conditions, with particle aerodynamic diameter up to 60 μm, wind velocities up to 8 m/sec and breathing patterns corresponding to most conditions of physical work. Vincent and Mark [1982], in Edinburgh, United Kingdom, worked with a model head mounted on a full-size torso in a large open-jet wind tunnel (Figure 2), extending the range of particle aerodynamic diameter up to 120 μm.

All of this research was conducted from the viewpoint that the human head is, in fact, a blunt dust sampling probe. However, unlike most dust sampling probes built and used for the purpose of dust measurement, dust sampling conditions for the head are complicated by the periodic fluctuations imposed by the breathing pattern, the shape of the sampler

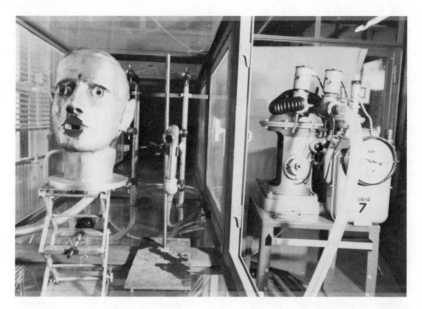

Figure 1. Experimental arrangement for determining inhalability in the Essen wind tunnel.

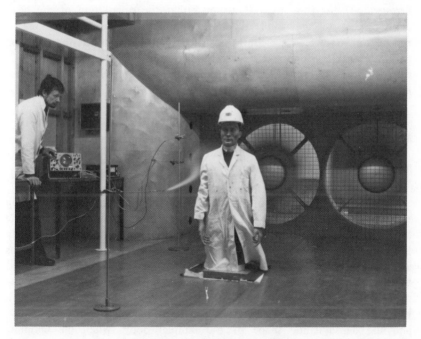

Figure 2. Edinburgh wind tunnel.

body and orifices, and the arbitrariness of the relative wind direction. Thus, it is not easy to construct a coherent theoretical model of the sampling process for the human head. Nevertheless we can say that inhalability, like the entry efficiency of "ordinary" dust samplers, is controlled by the inertial movement of the airborne particles in the complex airflow about the head and torso. Individual results for selected parameter combinations (Figure 3) show a wide range of values for the sampling efficiency of the head. Results like this may be shown to vary greatly from one experimental setup to another, depending, for example, on the presence (or otherwise) of the torso, wind tunnel blockage and analytical methods. However, when the results are averaged uniformly over all possible orientations of the head with respect to wind, this variability is substantially reduced. Now to a fair approximation, the experimental data for inhalability I can be plotted as a single function of particle aerodynamic diameter d, as shown in Figure 4 [Vincent and Armbruster 1981]. Here the continuous curve represents an eye-drawn "best fit" and the broken lines are the spread due to variations in wind speed. This curve is representative for humans in situations of normal physical work, and may be considered as a basis for a quantitative definition of inhalability. For particles with aerodynamic diameter up to about 40 μ, inhalability drops from 1 down to about 0.5, and remains more or less constant as far as 120 μm (which is the largest particle size investigated). As long as it is assumed that the orientation of the head with respect to the wind direction is uniformly distributed, there is no reason to expect a "cutoff" as suggested by the Ad hoc Working Group [1981]. From the experimental results like those in Figure 3, such a cutoff will only occur if the $0°$ orientation is grossly underrepresented in comparison to orientations between 90 and $180°$.

MEASUREMENT OF INHALABLE DUST

Theoretical Background

Vincent and Mark [1982] proposed a simple physical model to help explain the dust sampling characteristics of blunt probes (including the human head itself). They took into account the airflow pattern immediately in front of the probe, particularly disturbance of the homogeneous streamline pattern by the blunt body of the probe. They suggested that sampling efficiency is a function of the inertial parameter of the dust particles moving near the sampling orifice (the Stokes number), the ratio of the freestream and sampling air velocities, respectively, and the aerodynamic shape of the sampler body. When the sampling

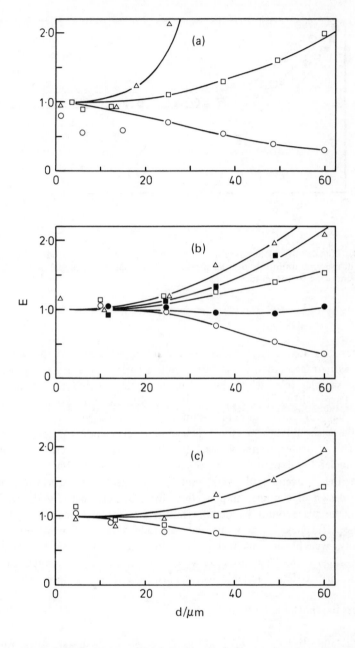

Figure 3.   Sampling efficiency E of the human head as a function of particle aerodynamic diameter d, facing the wind and breathing through the mouth: (a) subject
at rest, (b) normal work, and (c) hard work. (Wind speeds: o = 1; ● = 2; □ = 4; ■ = 6;
△ = 8 m/sec.)

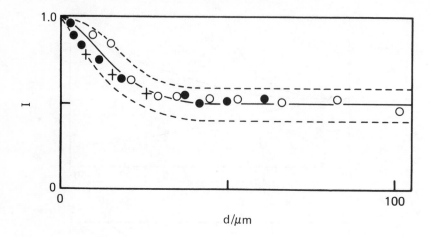

Figure 4. Inhalability of airborne dust I as a function of particle aerodynamic diameter d, averaged uniformly over all possible orientations of the model, normal work rate, and wind speeds up to 8 m/sec.

characteristics are integrated over all possible orientations of the sampling orifice with respect to the wind direction, a basis is obtained for designing a static sampler inlet whose entry characteristic matches the inhalability curve. Particularly of interest is the qualitative prediction from the model that for such a sampler—uniformly oriented with respect to the wind—there should be no cutoff in sampling efficiency for large particle sizes. This provides some physical justification for the results obtained in the inhalability experiments described in the previous section.

From the Vincent and Mark model, the main geometrical criterion for the design of a blunt sampling inlet for inhalable dust is that sampling should take place omnidirectionally in the horizontal plane. This ensures that the efficiency of sampling will be independent of wind direction, and only weakly dependent on wind speed and sampling air velocity. At this stage of understanding, however, the details of the actual sampler body and orifice shapes and dimensions must be determined empirically.

Samplers for Inhalable Dust

Ogden and Birkett [1978] proposed a static sampling inlet whose entry characteristics reasonably closely matched their limited inhalability data for particles with aerodynamic diameter up to about 30 μm [Ogden and Birkett 1977]. This is the Orb sampler, shown in Figure 5. It has

Figure 5.   The Orb inhalable dust sampling inlet.

a spherical shape, around which, at a determined latitude above the equator, are located equidistantly spaced circular sampling orifices; air is sampled at 2 liter/min. It is seen to meet the broad geometrical criteria we have specified. The flow pattern of moving air near the probe is modified by the presence of the cap (or halo), supposedly to make it resemble as closely as possible that about the human head. The entry efficiency of the Orb sampler inlet (Figure 6) falls somewhat below the inhalability curve taken from Figure 4, but it is nevertheless acceptably close for particles in the size range indicated. It is only weakly dependent on wind speed [Ogden and Birkett 1978].

Recent work with the Orb sampling inlet (along with other blunt dust samplers) has revealed that the inlet is prone to large oversampling errors due to particle blowoff from the external surfaces [Vincent and Gibson 1981]. This problem is particularly in evidence when sampling dry dusts that contain coarse particles greater than 30 μm in diameter.

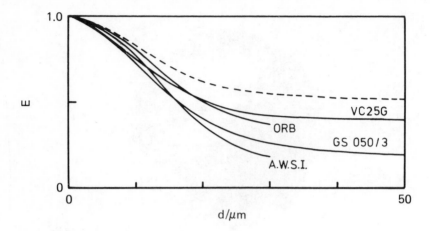

Figure 6. Sampling efficiency E as a function of particle aerodynamic diameter d for the ORB, Gravicon VC25G, GS 050/3 and All-Weather Sampling Inlet; shown in comparison to the inhalability curve taken from Figure 4.

Samplers Designed for "Total Dust"

In the Federal Republic of Germany, the need is recognized to specify certain physical sampling criteria when measuring "total dust." The MAK [1981] list states that sampling inlets are appropriate for collecting "total dust" if their sampling intake velocities are of the order of 1.25 m/sec. This applies to omni- and unidirectional sampling inlets. According to another convention [VDI 1980], inlets with unidirectional sampling orifices should be operated in the 45° position relative to the wind. At present, however, only samplers with annular horizontal slot openings, sampling omnidirectionally, are in use. Fortuitously, perhaps, they satisfy the main geometrical criterion we have suggested for an inhalable dust inlet. The sampling efficiencies of two such inlets were investigated to see how they perform in this respect.

For industrial workplaces (except in the mining industry), the Gravicon VC25G [Coenen 1973] is used, sampling at 400 liter/min (Figure 7). For the ambient atmospheric environment, the small filter unit GS 050/3 was developed [Laskus et al. 1980], sampling at 50 liter/min (Figure 8). The GS 050/3 sucks the dust-laden air from behind the annular slot in an upward direction, which causes losses by elutriation within the coarse particle fraction. There are no such losses with the VC36G, where the filter is located immediately below the annular slot.

Figure 7.   The "total dust" sampler Gravicon VC25G for use at workplaces.

The averaged sampling efficiency of the Gravicon VC25G for wind speeds in the range 1–8 m/sec lies very close to the inhalability curve (see Figure 6). When the individual results are considered, however, then it becomes obvious that there exists a strong dependence on wind speed (Figure 9). This was not the case for the human head, and not predicted by Vincent and Mark for such an omnidirectional sampling inlet.

Figure 8. The small filter unit GS 050/3 for "total dust" sampling in the ambient atmospheric air.

From the results, it could be argued that only at intermediate wind speeds would such a device be capable of satisfying the requirements for sampling inhalable dust. The ambient wind speeds prevailing at some workplaces would, however, come into this category, and so the Gravicon VC25G sampler could be used as an inhalable dust inlet in such situations. For the atmospheric environment, where it is relatively

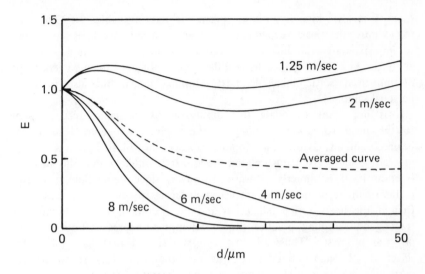

Figure 9.   Sampling efficiency E as a function of particle aerodynamic diameter d for the Gravicon VC25G at several wind speeds.

rare to find airborne particles larger than about 20 μm in aerodynamic diameter, the GS 050/3 would also be acceptable (see Figure 6).

The American All-Weather Sampling Inlet [McFarland et al. 1977] was designed to meet EPA sampling requirements. However, its geometrical configuration also happens to meet the criteria that we have specified for an inhalable dust inlet. We have not yet tested the entry characteristics of this instrument ourselves, but from the results published by its developers (reproduced also in Figure 6), its sampling efficiency also appears to fall somewhat below the inhalability curve.

DISCUSSION

Simple blunt dust sampler theory can be extended to show that it may be feasible to develop a family of static sampler inlets whose entry characteristics match up to the human inhalability curve that was established after the Fifth International Symposium on Inhaled Particles at Cardiff [Vincent and Armbruster 1981]. Sampling efficiency characteristics for a number of different inlets that meet the first basic geometrical requirement (omnidirectional sampling in the horizontal plane) appear in general to support this working hypothesis. From the data available, the Orb seems to exhibit the least dependence on wind speed. Work is continuing to improve understanding of the physics of blunt dust samplers and to apply the knowledge gained toward developing not only the next generation of improved static inhalable dust sampling inlets, but also improved personal samplers.

It is important to relate this discussion of inhalable dust sampling to the need for good sampling practice, which we referred to at the beginning of this chapter. In sampling to assess the risk to a given workforce of contracting a given dust-related disease, it is unlikely that the inhalable fraction itself is directly relevant. However, since deposition in a given region of the respiratory tract involves two successive selection processes—the first at the head during inhalation and the second inside the respiratory tract—good sampling practice should attempt to represent both of these physical processes. Thus, all dust samplers should first have inlets that select the inhalable fraction of the airborne dust. Any subfraction of particular health-related interest can be separated inside the body of the instrument. Technically, the latter should be the easier to achieve. In the sampling of respirable dust, the question is fairly academic since, for such fine particles, inhalability (as well as the entry efficiency of most sampling instruments) is always close to unity. The problem becomes more significant for coarser particles—for example, the TB subfraction—over whose aerodynamic size range inhalability varies substantially.

Finally some remarks on the question of "total dust" are appropriate. There seems to be no suitable single definition of "total dust"; rather, it becomes defined in practice by whatever means is used for sampling it. However, since the physical process of sampling, as governed by the detailed air and particle motions in the vicinity of any sampling device, is strongly dependent on its design and external factors, then it is not surprising that considerable variability exists in the sampling of "total dust" when different samplers are used. Therefore we consider that the concept of "total dust" as it stands at present is unsatisfactory. It is surely more sensible to start from the biologically relevant inhalable dust as the baseline for all dust sampling.

ACKNOWLEDGMENTS

The authors wish to thank the government of Nordrhein-Westfalen (West Germany), the British National Coal Board and the Commission of European Communities for their financial support of this work.

REFERENCES

Ad hoc Working Group to Technical Committee 146—Air Quality, International Standards Organization (1981) "Size Definitions for Particle Sampling," *Am. Ind. Hygiene J.* 42:A64-A68.

Armbruster, L., and H. Breuer (1982) "Investigations into Defining Inhalable Dust," in *Inhaled Particles V*, W. H. Walton (Oxford: Pergamon Press Ltd.), pp. 21-32.

Coenen, W. (1973) "Ein Neues Meßgerat zur Beurteilung Fibrogener Staube am Arbeitsplatz," *Staub-Reinhalt. Luft* 33:99.

Laskus, L., D. Bake, and L. Armbruster (1980) "Untersuchungen an Probenahmesystemen fur den Hygienisch Relevanten Schwebestaub im Staubkanal und in der Außenluft," *Staub-Reinhalt. Luft* 40:18-26.

MAK (1981) "Mitteilung der Senatskommission zur Prufung gesundheitsschadlicher Arbeitsstoffe," *Maximale Arbeitsplatzkonzentration* (Boppard, FRG: Harald Boldt Verlag).

McFarland, A. R., J. B. Wedding, and J. E. Cermak (1977) "Wind Tunnel Evaluation of a Modified Andersen Impactor and an All Weather Sampler Inlet," *Atmos. Environ.* 11:535-539.

Miller, J. M., D. E. Gardner, J. A. Graham, J. R. Lee, W. E. Wilson, and J. D. Bachmann (1979) "Size Considerations for Establishing a Standard for Inhalable Particles," *J. Air Poll. Control Assoc.* 29:610-615.

Ogden, T. L., and J. L. Birkett (1977) "The Human Head as a Dust Sampler," in *Inhaled Particles IV*, W. H. Walton, Ed. (Oxford: Pergamon Press Ltd.), pp. 93-105.

Ogden, T. L., and J. L. Birkett (1978) "An Inhalable Dust Sampler for Measuring the Hazard from Total Airborne Particulate," *Ann. Occup. Hygiene* 21:41-50.

VDI (1980) "Feststellen der Staubsituation am Arbeitsplatz zur gewerbehygienischen Beurteilung," Verein Deutscher Ingenieure Handbuch Reinhaltung der Luft, Bd. 4, VDI 2265, Berlin, FRG: Beuth-Verlag GmbH.

Vincent, J. H., and L. Armbruster (1981) "On the Quantitative Definition of the Inhalability of Airborne Dust," *Ann. Occup. Hygiene* 24:245-248.

Vincent, J. H., and H. Gibson (1981) "Sampling Errors in Blunt Samplers Arising from External Wall Losses," *Atmos. Environ.* 15:703-715.

Vincent, J. H., and D. Mark (1982) "Applications of Blunt Sampler Theory to the Definition and Measurement of Inhalable Dust," in *Inhaled Particles V*, W. H. Walton, Ed. (Oxford: Pergamon Press Ltd.), pp. 3-19.

CHAPTER 18

AEROSOL SAMPLING REQUIREMENTS TO ASSESS THE HEALTH RISK OF SPECIAL PARTICULATES

David L. Swift

Division of Environmental Health Engineering
School of Hygiene and Public Health
Johns Hopkins University
Baltimore, Maryland

ABSTRACT

Airborne particulates that produce pneumoconiosis historically have been sampled using instruments whose acceptance characteristic conforms to the British Occupational Health Society (BOHS), Atomic Energy Commission (AEC) or American Conference of Government Industrial Hygienists (ACGIH) curves. In this sense, the particles capable of reaching the nonciliated regions of the respiratory tract are clearly defined as "respirable." The concept was also applied by the AEC to radioactive particles that could remain in the lungs for long periods of time and result in a significant dose.

More recently, an acceptance curve describing "inhalable" particles has been proposed with the intention of defining airborne particulates capable of penetrating (under normal inspiratory breathing patterns) beyond the larynx, thus reaching either the tracheobronchial (TB) airways or the nonciliated regions. This concept has not, as yet, been applied to specific industrial exposure situations, nor have workplace samplers been designed, although a sampler for the outdoor environment based on this criterion has been designed and tested.

No sampler acceptance curves have been proposed to assess the health risk for particulates whose site or mode action does not fit the above criteria or whose deposition characteristics are not appropriately matched to the sampling condition. Special particulates of interest in the occupational environment include cotton dust, wood dust, asbestos and other fibers. The site of action of wood dust is the nasal passage; particulates that pass beyond the passage or that do not enter do not pose a known health risk. A sampler for health risk should be based on an acceptance curve derived from deposition studies such as those carried out in our laboratory. The deposition of fibers in the upper airways has been considered theoretically, but no experiments have been performed to confirm these calculations.

Other particulates which should receive special attention in terms of sampling include soluble or hygroscopic particulates which may be either solid or liquid. The growth of such particulates in the respiratory tract is not simulated in sampling instruments.

INTRODUCTION

There are a number of reasons for sampling airborne particles in an occupational environment. One may wish, initially, simply to characterize the aerosol spatially or temporally or with respect to its physical or chemical characteristics. If some process or procedural change is undertaken, the influence of this change on aerosol character may be desired. The effect of control technology on the concentration or character of the aerosol may be desired. A comparison of aerosol concentration to some standard imposed by a process requirement (e.g. "clean room" conditions for microelectronics) or to protect from dust explosion may be desired. The dynamics of aerosols (coagulation, surface deposition, etc.) may be of interest or the recovery of a valuable product may be desired.

However, in most instances, the object of aerosol sampling is to estimate worker exposure to aerosol via the respiratory tract and, from this estimate, to obtain an idea of the health risk associated with this exposure. Most of the aerosol standards for the work environment are established to protect health and are based on the assumption that a worker can be exposed to an airborne particulate 40 hr/week for a working lifetime of up to 50 years. It is possible that aerosols may produce health effects involving a different route of entry or site, such as eye irritation or dermal effects, but in most cases with aerosols, the respiratory route is of primary importance.

Not all of the aerosol particles present in an occupational environment may represent a specific health risk, and in a control strategy it is important to know what part of the aerosol is associated with the health risk, so that effort and expense will not be expended with little or no reduction in the occurrence of disease related to aerosol exposure. The same may be and probably is true for nonoccupational indoor or outdoor exposure to aerosol particles, but in this case the process of establishing relationships between specific particle exposures and respiratory or other diseases is more difficult than in the occupational setting, where the concentration is higher and the exposure is usually more clearly defined.

SAMPLING FOR "RESPIRABLE" PARTICLES

Although one might logically think that "respirable" particles are those that can enter the respiratory tract during normal breathing, the concept of respirable particles as those particles able to reach the nonciliated regions of the respiratory tract is the oldest concept that defines and is related to sampling methods.

In his review of the basis for respirable sampling, Lippmann [1978] points out the rationale of the two primary definitions of respirable particles and compares these definitions to respiratory tract deposition data. The first definition and method proposed by the British Medical Research Council (BMRC) in 1952 was aimed at predicting the risk of pneumoconiosis in its several forms and was adopted in 1959 by the Johannesburg International Conference on Pneumoconiosis [Orenstein 1959].

The other definition of respirable particles proposed by the AEC in 1961 was intended to include insoluble particles that deposited in the nonciliated region of the respiratory tract and remained for long periods of time. In this case, particles containing long-half-life isotopes were of primary interest, so the risk of damage to the lungs and other organs to which such insoluble particles might be carried was to be estimated by such sampling procedures. In this case, data on human lung deposition [Brown et al. 1950] were used to derive the cutoff efficiencies for different particle sizes. Sampler efficiency as a function of particle diameter for these definitions is shown in Table I. ACGIH adopted a definition of respirable particles almost identical with the AEC definition (except that for 2 μm the acceptance is 90% rather than 100%) and, like BMRC, has applied this definition to the class of insoluble dust that produces pneumoconiosis.

In 1966 the International Commission on Radiation Protection (ICRP) Task Group on Lung Dynamics published a model for aerosol deposition and retention in the human respiratory tract [Task Group on Lung Dynamics 1966]. The intention was to supply a model that could be used to calculate dose for internal emitters of ionizing radiation. Curves for percent aerosol deposition of unit density spheres (for particles >0.5 μm, aerodynamic equivalent diameter) for three breathing conditions

Table I. Acceptance Curve for BMRC, AEC, ACGIH Respirability Criteria

Percent Passing Precollector	Particle Diameter (μm)		
	BMRC	AEC	ACGIH
0	7.1	10	10
25	6.1	5	5
50	5.0	3.5	3.5
75	3.5	2.5	2.5
90	2.2		2.0
100		2.0	

(rest, light work, heavy work) were presented for three zones of the respiratory tract (nasopharyngeal, TB and pulmonary) based on deposition data published to that date.

The percent pulmonary deposition for nasal breathing at rest presented by the ICRP group is plotted in Figure 1 with the acceptance curves of the BMRC and ACGIH and alveolar deposition data for oral and nasal breathing carried out at New York University. The sampler acceptance curves have upper size cutoffs for respirable particles that match the experimental data better than the ICRP curve, which predicts as much as 10% of the 20-μm-diameter particles reaching the nonciliated region at this breathing condition.

According to the data, there is considerably more mass reaching the nonciliated area in mouth breathing compared to nasal breathing. The area under the mouth curve plotted as aerosol mass is more than three times the area under the nose curve. This means that the respirable function sampled for a given situation is an estimate of the risk of pulmonary disease (proportional to the quantity of aerosol delivered to the nonciliated region) for a given mode of breathing; for the same mass sampled, the risk is higher if the mode of breathing permits more deep-lung deposition. This presumably occurs at a given work level where nasal breathing is no longer adequate for respiratory demand. Of course, individual variation in deposition for a variety of reasons, including preexisting lung disease, also alters the risk, but it is difficult to include such factors in a sampling strategy or definition.

SAMPLING FOR "INHALABLE" PARTICLES

Motivation for the definition of a newer standard and sampling method for airborne particles came from the U.S. Environmental Protection

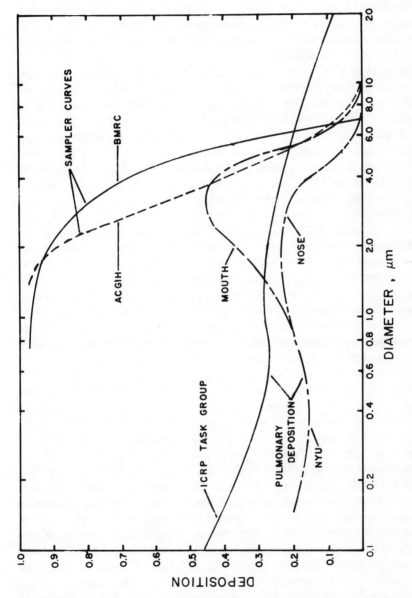

Figure 1.   Sampler curves and pulmonary deposition data [Lippmann 1978].

Agency (EPA) in their consideration of criteria for the revision of the Clean Air Act. Previous sampling of outdoor air particulates has been carried out throughout the United States, employing the high-volume sampler. This method has been used for many years, despite criticism that the sampling method was wind-direction-sensitive and that its upper particle diameter collection characteristics included material that might not represent a significant human health risk.

Miller et al. [1979] proposed the concept of inhalable particles as those particles that can traverse the oral or nasal passage past the larynx and enter the TB tree. The existing data for deposition of monodisperse aerosols in the nasal and oral passage were reviewed and, on the basis of these data, a proposal for 50% cutpoint of 15 μm aerodynamic equivalent diameter was proposed, such that particles smaller than this size would be denoted inhalable. It was proposed that a sampling instrument and procedure be developed to permit a mass or other analysis of outdoor airborne inhalable particulates to be performed.

On the basis of aerosol chemistry and the source of particulates, a second cutpoint of 2.5 μm equivalent diameter was proposed to separate inhalable particles into coarse and fine fractions. This cutpoint has no physiological basis, from neither aerosol deposition nor epidemiological studies, but was proposed purely on the basis of outdoor aerosol chemistry. There is no basis, at present, for adopting such a cutpoint for occupational aerosol sampling, since there is no evidence that the physical or chemical characteristics of aerosol sampled in the occupational setting undergo any change at this particle size, nor does aerosol deposition in the respiratory tract show a discontinuity or minimum at this particle diameter.

For outdoor air particulates, an instrument that meets the 15-μm cutoff criteria has been designed and tested [Wedding 1981], and its revised form has been found to meet the cutoff criteria relatively independently of wind speed as shown in wind tunnel studies [McFarland et al. 1977]. Such an instrument for sampling in an occupational environment with an appropriate sampling rate has not yet been designed.

Unfortunately, the term "inhalable dust" has also been used by Ogden and Birkett [1978] to refer to dust in the occupational environment that enters the respiratory tract itself, not the TB tree. A sampling instrument to estimate the quantity of suspended particulate entering the nasal or oral passage, giving particular attention to the effect of wind speed and orientation on the sampling efficiency, was designed by these authors. Aerosol is sampled into the instrument at 2 liter/min and collected on a membrane filter for subsequent gravimetric, microscopic or other analysis. The collection characteristics of the instrument are based on studies of sampling efficiencies of a "head dummy" in a wind tunnel breathing (by mouth or nasal passage) particles up to 30 μm [Ogden and Birkett 1978].

INSPIRABLE DUST SAMPLING

To avoid the confusion about inhalable dust discussed above, it has been proposed that aerosol particles entering the nasal or oral passage during breathing be denoted "inspirable aerosol." Although this definition has not been formally adopted, the proposed curve of the inspirable fraction has a 50% cutpoint at 30 μm equivalent diameter with a selection curve shown in Table II.

Particles that enter the respiratory tract according to this selection curve could be deposited in any of the regions of the respiratory tract, and if the health risk estimate is based on sample mass meeting this standard, it would be assumed that particles would have the same effect whatever region they entered or were retained at. This assumption does not seem sensible from an effects viewpoint, particularly when considering particulates that have a special health risk associated with a specific respiratory region.

For soluble particles that can enter systemic circulation from any respiratory compartment, the inspirable aerosol mass should give a good relative estimate of the health risk of inhalation of a particular dust substance. Insoluble dusts that produce irritation, are cytotoxic to specific cells of the respiratory epithelium or lead to macrophage dysfunction will produce effects, depending on whether the majority of the particle mass is deposited in the nasopharyngeal, TB or pulmonary compartment.

The primary utility of this concept of inspirable particles is to exclude from consideration very large particles, which cannot produce effects in the respiratory tract because they are not able to enter the nasal or oral passage. If such large particles are excluded from collection, the particles

Table II. Acceptance Curve for Inspirable Aerosol Definition

Percent Passing	Particle Diameter (μm)
95.6	1
85.8	4
76.8	8
70.3	12
60.6	20
51.7	30
39.2	50
30.1	70
19.7	100
9.2	140
1.0	180

collected will give a relative impression of the overall respiratory tract delivery of aerosols. However, it is well known that particles smaller than ~4 μm are not removed quantitatively when breathed "normally" by "normal" individuals. This tendency to exhale significant fractions of small nonhygroscopic particles reaches its maximum at about 0.5 μm, where at normal breathing conditions approximately 10% deposition occurs in the nonciliated region.

Consider two aerosols of the same substance and mass concentrations. If one has most of its mass surrounding 8.0 μm and the other most of its mass surrounding 0.5 μm, the first will have 76.8% collected in the "inspirable aerosol sampler," whereas the second will have ~100% collected. However, in human inhalation, approximately 100% of the inspired large aerosol will be deposited in the upper and TB airways, while approximately 10–20% of the small aerosol will deposit only in the nonciliated region. The relative health risk of the two aerosols is not likely to be in proportion to the mass collected by the inspirable aerosol sampler.

If inspirable dust could be split into three fractions according to the region of the respiratory tract that received most of the aerosol mass, the relative mass of aerosol in each fraction would be related more directly to the health risk for a specific dust species. A sampling instrument that accomplishes this fractionation does not exist, but there is no fundamental reason why such a fractionator could not be constructed according to the present measurement of regional aerosol deposition in normal adult humans. For special particles (to be discussed below), the mass in one or more fractions may represent the relative health risk, since the effect may only be felt in a single respiratory compartment (other than the nonciliated region) to which respirable particles are accessible.

WOOD DUST PARTICLES

Although several special species of wood are known to produce allergic reactions, interest in the occupational exposure of workers to wood dust has heightened since it was first reported by Hadfield [1971] that woodworkers in the furniture industry had abnormally high rates of cancer of the nasal passage and sinuses.

For these cancers to appear in the upper respiratory airways, it is necessary for particles to enter the nasal passage and be deposited. Particles that enter the mouth or pass through the nose onto the more distal regions of the respiratory tract are not believed to present a risk of cancer. Thus, only the particles that enter and deposit on the nasal mucosa should be considered a potential health risk.

Thus, the particles to be excluded from sampling are those so large that they do not enter the nasal passage and those small enough to pass through the nasal airways during inhalation and exhalation. The upper cutoff for inspirable particles (Table II) is a curve for acceptance of the nasal passage based on the studies in wind tunnels by Vincent and Mark [in press] and Armbruster and Breuer [in press] employing mannequin heads. The acceptance of the nasal passage for large particles in still air was studied in vivo by Breysse and Swift [1980]. In this study, the 50% acceptance diameter was also approximately 30 μm, similar to the inspirable particle curve, but the decrease in acceptance at larger particle size was more marked. The extrapolated value for zero acceptance based on the experimental data was 40-50 μm, while for the inspirability curve, zero acceptance is at 185 μm.

Some of the discrepancy between these results is related to the increased acceptance of larger particles in the nasal passage at low wind speed (wind direction directly onto the face) which has been demonstrated in the wind tunnel studies. However, more in vivo studies are recommended for conditions of both mouth and nasal breathing in known wind flow conditions with large particles to resolve the issue of large particle acceptance. In the outdoor nonoccupational environment, this may not be particularly important, because so little of the mass of particles is found in particle sizes above 40 μm, but in certain occupational settings (e.g., wood dust exposure), the larger particles may account for a majority of the mass of particulates, and their accessibility to the upper airways is of crucial importance to the risk estimate of nasal or sinus disease.

The lower cutoff for nasal deposition is approximately 5 μm for normal nasal breathing, so a sampler to collect particulates related to nasal or sinus disease should not collect particles smaller than 5 μm and larger than 50 μm, if the in vivo data of Breysse and Swift are assumed correct. No such sampler is commercially available at present, but the design and construction of such a device does not seem to present any unusual difficulties.

COTTON DUST PARTICLES

In the United States, studies of the health effects of exposure to cotton dust have generally employed a sampling instrument known as vertical elutriator, originally designed to have an upper size cutoff at 15 μm. All particles smaller than this size are collected on an absolute filter for gravimetric analysis. It is not clear how the original decision to design the elutriator with a 15-μm cutoff was made, but apparently it was desired to collect what now would be described as inhalable dust.

There is a difference in theory between the collection characteristics of the vertical elutriator and the inhalable sampling curve: the vertical elutriator was intended to have an absolute cutoff at 15 μm, so that its acceptance curve is similar to the inhalability criterion.

The disease condition associated with cotton dust exposure is byssinosis, characterized initially by airway reactivity and chronic bronchitis, and in many cases leading eventually to chronic obstructive pulmonary disease (COPD). It would appear that cotton dust particles depositing either in the bronchial airways or in the nonciliated (pulmonary) region would be capable of producing the response observed, but bronchial deposition, especially in the medium and large airways, is probably more important, leading to the acute bronchitis and observed changes in mechanical properties of the lungs.

When bronchial airways become constricted, either acutely or chronically, particles that otherwise would easily penetrate to the pulmonary region can be deposited in the bronchial airways and give a high local surface concentration. Once the symptoms of cotton dust disease begin to appear, all particles that enter the tracheobronchial airways can be deposited in this region. Thus, it would seem reasonable to use the inhalability criterion as a measure of risk for cotton dust disease.

Is there a possibility that particles deposited in the upper airways (nasal or oral passage) may be important for cotton dust disease? This does not seem likely, although it has been suggested by some investigators that acute bronchial response can be elicited by particle deposition in the nasal passage. The response from direct bronchial deposition of particles is probably much more potent, in that acute constriction, reduced mucociliary clearance and other chronic manifestations of irritancy can be produced. However, the role of upper airway deposition in the acute manifestation of cotton dust exposure should be investigated so that the use of inhalable sampling can be on a firmer ground.

ASBESTOS FIBERS

There are three chronic disease conditions ordinarily associated with exposure to asbestos fibers: asbestosis (a type of pneumoconiosis), bronchial cancer and pleural mesothelioma. In each case, it is necessary for the fiber to be deposited in the respiratory tract in such a way that it has long-term access to the appropriate tissue to produce the appropriate lesion.

Asbestosis, occurring in the pulmonary spaces, is undoubtedly related to pulmonary deposition and long-term retention in this region. Likewise, it has been hypothesized that pleural mesothelioma results from the deposition of fibers in the pulmonary region and their subsequent migration to the pleural spaces. For these two diseases, risk most likely depends on the delivery of fibers to the pulmonary region.

Since pulmonary deposition of fibers is related to disease risk for these conditions, one may ask whether classical respirable sampling techniques are appropriate. At present, the penetration of fibers through the nasal and bronchial airways is not known as a function of fiber length and diameter, so that the simple concept of aerodynamic diameter cannot be employed as it is for compact dust particles. It is likely that long fibers, which have been considered more hazardous, can be removed in the nasal or bronchial airways to some degree by direct interception. Although this has been predicted theoretically by Timbrell and Harris [1977], no measurements have been reported to confirm these calculations, which would provide a basis for sampling.

Furthermore, the question of regional deposition of fibers and the appearance of bronchial cancer is yet unanswered. Since it appears much more frequently in smokers, the increased bronchial deposition and reduced mucociliary clearance associated with chronic smoking and attendant bronchitis suggest that fibers capable of reaching middle bronchi can play a role in this disease. This suggests that an inhalable criterion should be used for asbestos fibers: sampling fibers that are capable of penetrating beyond the nasal or oral passage. Again, the penetration curve of the nasal or oral passage for fibers is not known experimentally; thus, this information is needed before a suitable sampling instrument can be devised.

ULTRAFINE PARTICLES

Ultrafine particles may arise from a number of situations in which chemical or physical conversion from gas to particle phase takes place. Such particles have an effective diameter less than 0.1 μm and may be unitary or agglomerated by coagulation. The deposition characteristics of such particles are not well known experimentally; the few experiments that have been performed for particles having diameter ranging from 0.02 to 0.1 μm [Swift et al. 1977] suggest that deposition occurs primarily in the nonciliated region, but that some deposition of the smaller sizes takes place in the nasal airway. These particles may have very large surface/volume ratios and thus may carry adsorbed species into the nonciliated region. The mass contribution of these particles under most circumstances is small, but their effects may be related more to surface area than mass. For this reason, the number or surface area of particles in this size range should be determined, rather than the mass. At present, instruments for continuously determining the number of ultrafine particles, such as condensation nuclei counters, are nonportable and complex in operation.

If such an aerosol is sampled by filtration for subsequent microscopic analysis, the filter must collect the particles with 100% efficiency and in such a manner that they can be visualized easily when prepared for electron microscopy. Techniques for sampling ultrafine particles need further development before they can be used routinely for health effect assessment.

HYGROSCOPIC PARTICLES

Sampling instruments for collecting respirable, inhalable or inspirable particles do not simulate the human respiratory tract with respect to the heat and moisture conditions at entrance. This may be an important consideration for hygroscopic particles where significant growth and upper respiratory deposition takes place that reduces the quantity of aerosol accessible to the nonciliated region.

Several studies in humans suggest that changes in deposition patterns do take place with hygroscopic aerosols, but there are inadequate data on the degree of growth and the temperature conditions in the airways under different breathing conditions. A sampling instrument that first exposes particles to air saturated with water vapor at 37°C would be a useful instrument to see if the respirable fraction changed markedly from the control situation.

REFERENCES

Armbruster, L., and H. Breuer (in press) "Investigations into Defining Inhalable Dust," in *Inhaled Particles V* (Oxford: Pergamon Press Ltd.).

Breysse, P., and D. Swift (1980) "Deposition of Large Particles in the Human Nasal Passage," paper presented at AIHA Meeting, Houston, TX, May 1980.

Brown, J. H., K. M. Cook, F. G. Ney, and T. Hatch (1950) "Influence of Particle Size upon the Retention of Particulate Matter in the Human Lung," *Am. J. Public Health* 40:450.

Hadfield, E. H. (1971) "Damage to the Human Nasal Mucosa by Wood Dust," in *Inhaled Particles IV* (Old Woking, UK: Unwin Bros.).

Lippmann, M. (1978) "Respirable Dust Sampling," in *Air Sampling Instruments* (Cincinnati, OH: American Conference of Government Industrial Hygienists).

McFarland, A. R., J. B. Wedding, and J. E. Cermak (1977) "Wind Tunnel Evaluation of a Modified Anderson Impactor," *Atmos. Environ.* 11:535.

Miller, F. J., D. E. Gardner, J. A. Graham, R. E. Lee, W. E. Wilson, and J. D. Buchmann (1979) "Size Considerations for Establishing a Standard for Inhalable Particles," *J. Air Poll. Control Assoc.* 29:610.

Ogden, T., and J. Birkett (1978) "An Inhalable Dust Sampler for Measuring the Hazard from Total Airborne Particulate," *Ann. Occ. Hygiene* 21:41.

Orenstein, A. J. (1959) *Proceedings of the Pneumoconiosis Conference, Johannesburg* (London: Churchill).

Swift, D. L., F. Shanty, and J. T. O'Neill (1977) "Human Respiratory Tract Deposition of Nuclei Particles and Health Implications," in *Airborne Reactivity* (LaGrange Park, IL: American Nuclear Society).

Task Group on Lung Dynamics (1966) "Deposition and Retention Models for Internal Dosimetry of the Human Respiratory Tract," *Health Phys.* 12:173.

Timbrell, V., and R. Harris (1977) "The Influence of Fiber Shape in Lung Deposition-Mathematical Estimates," in *Inhaled Particles IV* (Oxford: Pergamon Press Ltd.).

Vincent, J., and D. Mark (in press) "Application of Blunt Sampler Theory to the Definition and Measurement of Inhalable Dust," in *Inhaled Particles V* (Oxford: Pergamon Press Ltd.).

Wedding, J. (1981) "An Inlet for Inhalable Particulate Matter," EPA Contract Report, Colorado State University, Department of Mechanical Engineering.

PART 3

SAMPLING STRATEGY, DATA ANALYSIS AND INLET EFFICIENCY

CHAPTER 19

SAMPLING PLAN AND STATISTICAL EVALUATION FOR BITUMINOUS COAL RESEARCH UNDERGROUND DUST-SAMPLING PROGRAMS

G. D. Andria, R. D. Saltsman, J. A. Kost and J. L. Zalar

Bituminous Coal Research, Inc.
Monroeville, Pennsylvania

ABSTRACT

This chapter discusses recent sampling plans and statistical evaluations for Bituminous Coal Research (BCR) underground dust-sampling programs, all sponsored by the Bureau of Mines (BOM). The first program is concerned with the evaluation of the reduction of airborne dust due to the addition of a wetting agent to the water used for dust control. The sampling program primarily involved the use of half-shift gravimetric dust samples. The second program was an underground evaluation of a twin-scrubber dust collector mounted on a Jeffrey 120L continuous miner. The sampling program relied almost exclusively on the Simslin sampler. The use of the Simslin instantaneous dust monitor on a longwall face is also discussed. Finally, a proposed program to test respirable dust-control strategies (techniques and equipment) applicable to double-drum longwall shearers is discussed. The task of these projects is to define the relationship of various mine and environmental parameters to respirable-dust control and productivity levels. In general, evaluation of a specific dust-control measure is accomplished by conducting comparative tests, i.e., alternating the test sequence with and without the control measure under evaluation. Statistical sampling plans include practical considerations which limit how often a control is taken into and out of operation.

EVALUATION OF WETTING AGENTS

The purpose of this sampling program was to determine whether any reduction in airborne dust results from the addition of a wetting agent to the water used for dust control in underground coal mines [BCR 1981a]. Underground testing was conducted at the Mears Coal Company, Dixon Run No. 3 Mine in Indiana County, Pennsylvania, during August and September 1980. This single-section mine is located in the Lower Kittanning seam, which averaged approximately 46 in. thick during the tests. As shown in Figure 1, the mining system consisted of a Wilcox Mark 20 auger miner equipped with a bridge conveyor and belt conveyor for section haulage. The Mark 20 auger dust-suppression system includes wet augers, Conflow venturi sprays and a spray manifold.

Four wetting agents were evaluated during the underground tests: Surfynol 465, Aerosol MA-80, DC-13 and Dustallay. The choice of these four wetting agents and their concentrations was based on laboratory tests of coal wettability and the coal operators' experience. The concentration of Surfynol 465 and Aerosol MA-80 was maintained at 0.7%, while a concentration of

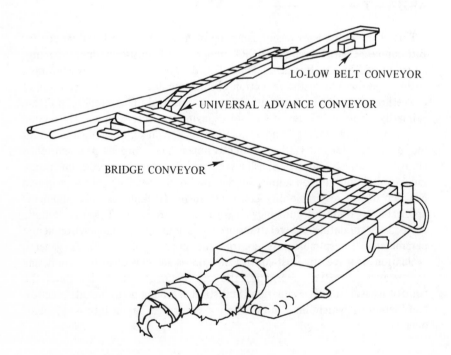

Figure 1.   Wilcox Mark 20 auger miner with coal transport system.

0.2% was used for DC-13 and Dustallay. Thus, the underground tests involved the use of four wetting agents and the plain mine water.

Three types of dust samplers were used to measure dust levels:

1. MSA Model G personal sampler with pulsation dampener and a 10-mm nylon cyclone, operated at 2.0 liter/min;
2. MSA Model G personal sampler with UNICO filter cassette and without a cyclone, operated at 2.0 liter/min; and
3. Simslin II instantaneous respirable-dust monitor with a single-channel horizontal elutriator operated at 0.625 liter/min.

The MSA Model G personal samplers with cyclones and the Simslin provided a measure of respirable dust, while the MSA samplers without cyclones provided a measure of the total airborne dust.

The general section layout and sampling locations for the underground tests are shown in Figure 2. All samplers were oriented with inlets directed into the

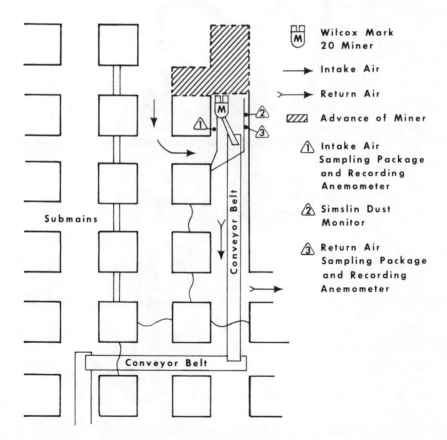

Figure 2.   General section layout and sampling locations for underground tests.

airflow approximately 18 in. from the floor. The sampling package (Figure 3) used in intake air consisted of two MSA personal samplers with 10-mm nylon cyclones. The samplers were placed behind the blowing brattice and maintained approximately 20 ft from the face. The return air sampling package (Figure 4) consisted of two MSA personal samplers with 10-mm cyclones and two MSA personal samplers with UNICO cassettes (total airborne dust samples). The return-air package was located behind the exhaust brattice approximately 25 ft from the face. Duplicate samples were collected for both the intake and return packages to provide a check on the proper functioning of the samplers by comparing the dust weights collected by

Figure 3.   Sampling package used in intake air to face.

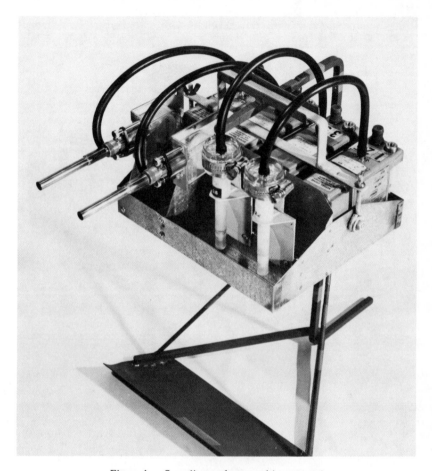

Figure 4.   Sampling package used in return air.

each unit. In addition, duplicate samples ensured that data would be available from each sampling location even if one sampler failed.

The Simslin instantaneous dust monitor (Figure 5) was placed, as a single unit, behind the exhaust brattice just upstream of the return-air sampling package. The Simslin dust monitor continuously monitors the respirable dust level and stores the information in a memory unit to be recalled at a later time. A permanent strip-chart record, as well as a gravimetric reading, is obtained for each shift.

Sampling was conducted for 22 shifts. All shifts were divided into two sampling periods, prelunch and postlunch. Thus, the 22 shifts provided 44

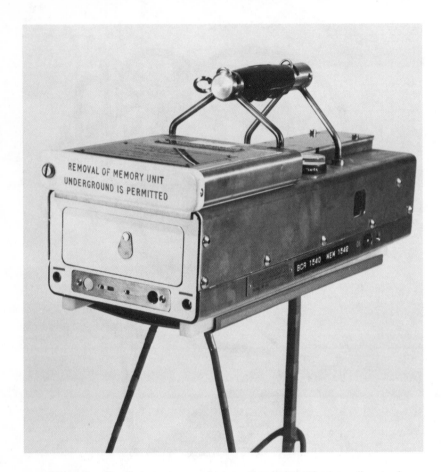

Figure 5.   Simslin instantaneous dust monitor positioned on sampling stand.

test periods. Due to various problems encountered during the actual sampling, 20 of the test periods were invalidated. Dust levels, as well as supporting operating data, for the remaining 24 valid test periods were used in the final analysis. Table I presents the actual test sequence used to evaluate the four wetting agents.

Since differences in airflow and production could contribute to variations in dust levels, the data were corrected for both airflow and tonnage. Two approaches were taken to adjust the data. The first was to calculate the average production and airflow for the 24 test periods and adjust individual dust levels to these values. The second method was to calculate a "normalized

Table I. Testing Sequence and Conditions[a]

Shift	Prelunch	Postlunch
1	I	I
2	I	I
3	I	I
4	I	I
5	I	I
6	I	I
7	E	A
8	E	A
9	E	B
10	C	B
11	I	C
12	A	E
13	I	I
14	E	I
15	E	E
16	I	E
17	E	I
18	E	D
19	D	D
20	B	D
21	I	I
22	B	D

[a]Codes for test conditions: A = Surfynol 465, B = Dustallay, C = MA-80, D = DC-13, E = water alone, I = invalid test.

dust emission" (mg/ton), which corrects the dust levels to a per-ton value based on sampling time, return airflow and production.

The data analysis involved two approaches. The first was to compare data from the test periods with plain water to those from all 14 test periods using a wetting agent. In other words, regardless of the wetting agent used, was there any difference between results with plain water and those with water and a wetting agent? The second approach was to compare the data with each of the wetting agents to those with plain water and then rank the four wetting agents.

Statistical comparisons between results with wetting agents and those with plain water conditions were made using the student's t statistic [Cox and Snell 1981]. Table II displays the percent dust reductions obtained through the use of wetting agents as well as the statistical level of confidence associated with each observed dust-reduction value. These results are presented for raw dust concentrations, adjusted dust concentrations, and normalized dust emissions based on both respirable and total airborne dust measurements.

Table II. Percent Dust Reductions Obtained Through the Use of Wetting Agents

	Dust Concentration (mg/m³)		Adjusted Dust Concentration (mg/m³)		Normalized Dust Emission (mg/ton)	
	Dust Reduction (%)	Confidence Level[a] (%)	Dust Reduction (%)	Confidence Level[a] (%)	Dust Reduction (%)	Confidence Level[a] (%)
Respirable, Personal						
All wetting agents	10		29	80	21	90
DC-13	+20[b]	50	20	50	23	70
Dustallay	10		30	70	24	70
Surfynol 465	35	70	29	60	18	50
Aerosol MA-80	33	60	46	70	25	60
Respirable, Simslin						
All wetting agents	17	60	33	99	25	95
DC-13	+11		24	60	28	80
Dustallay	22	60	40	80	22	70
Surfynol 465	44	80	35	70	24	70
Aerosol MA-80	38	70	43	70	26	60
Total Airborne						
All wetting agents	21	60	39	95	33	95
DC-13	+ 7		32	70	34	70
Dustallay	20	50	41	80	35	70
Surfynol 465	46	80	40	70	32	60
Aerosol MA-80	44	60	51	70	36	60

[a]Confidence level determined from student's t statistic.
[b]"+" indicates increase in dust level with surfactant.

Based on the test results obtained, the use of wetting agents appears to result in lower respirable and total airborne dust levels. The reduction in respirable dust amounted to approximately 27%, with 36% reduction in total airborne dust. As shown in Table II, similar dust reductions were obtained for each wetting agent tested, so a meaningful ranking of their effectiveness was not possible.

EVALUATION OF A MACHINE-MOUNTED TWIN-SCRUBBER UNIT

Underground testing and evaluation of the twin-scrubber unit was conducted at the Eastover Mining Company's Highsplint Mine in Harlan County, Kentucky [BCR 1981b]. The mine is located in the Harlan seam, which ranged 4–5 ft in thickness, including a 1-ft band of rock located near the top of the seam. The twin-scrubber unit was installed underground on a Jeffrey 120L continuous miner. Figures 6 and 7 show the left-side scrubber unit installed on the miner, including the scrubber inlet, water sprays, panel filter, 2500-ft^3/min centrifugal fan and the brattice exhaust duct.

Two types of respirable dust samplers were used for the underground

Figure 6. Overall view of left-side scrubber installation.

Figure 7.   Left-side scrubber installation.

evaluation: (1) MSA personal sampler with 10-mm nylon cyclone and (2) Simslin II instantaneous dust monitor with single-channel horizontal elutriator. Respirable airborne dust samples were collected at the miner operator's station and behind the exhaust brattice to determine the scrubber's effectiveness in reducing respirable dust levels. In addition, other recording instrumentation used during the tests provided continuous records of water flow, water pressure and airflow behind the exhaust brattice. Because this type of data was available for each cut, the data were analyzed on a per-cut basis. Seven sampling shifts resulted in 25 valid test cuts: 17 cuts with the scrubber on and 8 cuts with the scrubber off.

Table III summarizes the data collected during the underground evaluation. It shows that respirable dust levels behind the exhaust brattice and adjacent to the miner operator were reduced 88 and 76%, respectively, by using the scrubber. Statistical comparisons were made using the student's t statistic. The confidence levels for the reduction behind the brattice and at the miner-operator's location were 99.9 and 99%, respectively. Although average production and return airflow were very similar for both testing conditions, i.e., with and without scrubber, the dust concentration values behind the exhaust brattice were normalized for these operating variables; the normalized dust levels were reduced 85% at a 99.9% confidence level with the scrubber.

Table III. Summary of Data Collected Underground for
Evaluation of Twin-Scrubber Unit

	With Scrubber	Without Scrubber
Number of Test Cuts	17	8
Production (tons)	51	47
Water Flow (gal/min)	21	16
Flow Pressure (psig)	205	235
Airflow Behind Brattice (ft^3/min)	4400	4100
Dust Concentration (mg/m^3)		
Simslin Behind Brattice	7.0	56.8
Percent Reduction	88	
Simslin Adjacent to Miner Operator	1.4	4.5
Percent Reduction	76	
Normalized Dust Emission (mg/ton)		
Simslin Behind Brattice	780	5274
Percent Reduction	85	

Figures 8 and 9 show graphically the respirable-dust concentrations, measured behind the exhaust brattice and adjacent to the miner operator, for each test cut; they indicate clearly how effective the twin-scrubber system was in reducing respirable dust. Figures 10 and 11 show the Simslin traces from behind the brattice along with the continuous records of water flow and airflow for two individual cuts: first without the scrubber, and then with the scrubber on. The difference in dust levels is significant.

In addition to the Simslin dust monitor located between the miner operator and the machine conveyor, a personal sampler was worn by the operator. The respirable-dust concentrations as measured by both samplers were very comparable after correcting the personal sampler concentration to the Mining Research Establishment (MRE) equivalent. Based on the average dust concentration per cut as measured by the Simslin monitor, calculations were made to determine the maximum number of cuts that could be made in one shift before exceeding the dust standard of 2.0 mg/m^3. Without the scrubber, three cuts were possible before exceeding 2.0 mg/m^3; with the scrubber, 18 cuts could theoretically be made before the full-shift dust concentration at the miner operator exceeded 2.0 mg/m^3. A better indication of the scrubber's effectiveness is seen when the dust concentrations are compared for the same number of cuts. For example, if a section averaged eight cuts each shift, the dust concentration at the miner operator would be 1.0 mg/m^3 with the scrubber and 4.0 mg/m^3 without the scrubber.

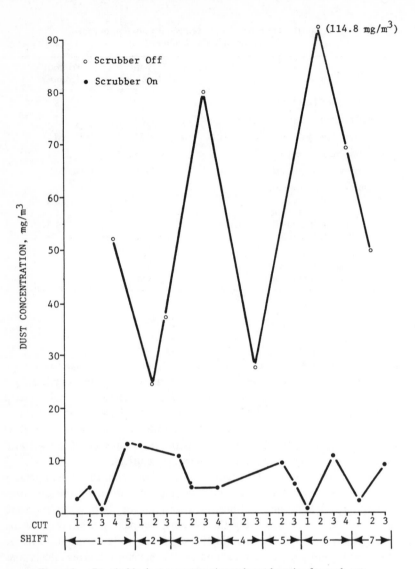

Figure 8.   Respirable dust concentration exhaust brattice for each cut.

In addition to the reduction in respirable-dust levels caused by the twin-scrubber unit, several members of the section crew, including the miner operator, operator helper and shuttle car operators, were very enthusiastic about the scrubbers and, in particular, about the increase in face visibility. Better visibility at the face creates a safer working environment and should result in a more productive mine.

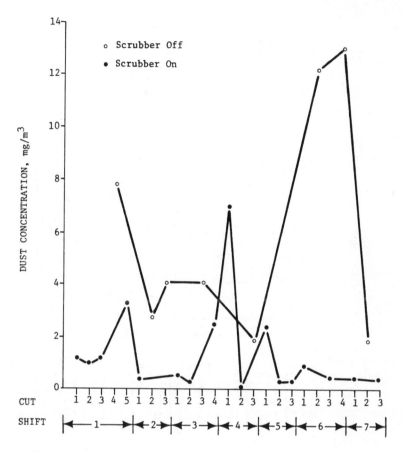

Figure 9.   Respirable dust concentration to miner operator for each cut.

USE OF SIMSLIN INSTANTANEOUS DUST MONITOR

The instantaneous dust monitors have the potential of providing far more information than was ever possible with gravimetric samplers. Another example of the type of data collection possible is the adjacent dust profiles recently obtained on a longwall double-drum shearer face. The Simslins were located 350 ft from the headgate on a face that was 550 ft wide. One sampler was positioned in the walkway while the second was located over the panline (face conveyor) approximately 6–10 ft away (Figure 12).

Two typical profiles are shown in Figure 13. The average concentrations over approximately 35 min were 32.1 mg/m³ over the panline and 11.8 mg/m³ in the walkway. A 3:1 difference in concentration was not unusual. Overall, the two profiles appear to be radically different. For example, at

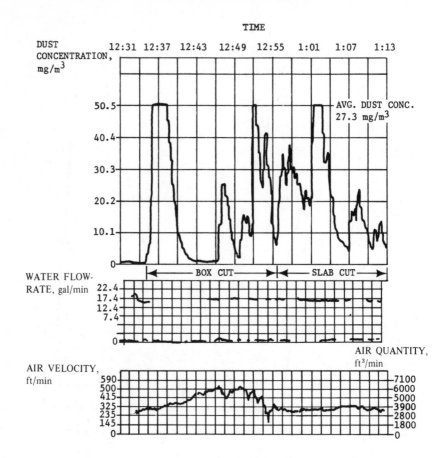

Figure 10.   Dust levels behind the brattice without the scrubber.

10:26 a.m., when concentrations in the panline approached 70 mg/m³, no concentration exceeded 18 mg/m³ in the walkway.

Perhaps the most surprising aspect is that these differences exist 150 feet downwind of the shearer. To date many researchers have assumed that, under turbulent flow, the dust cloud becomes well mixed and uniform in a relatively short distance. The problems presented by such large variations are immediately obvious. The selection of a sampling site is critical in any dust control program. For example, if a sampler is located in the walkway, a dust control method that could result in the elimination of the 70-mg/m³ peaks in the panline could go undetected. The question is: would there be a corresponding decrease in dust levels in the walkway?

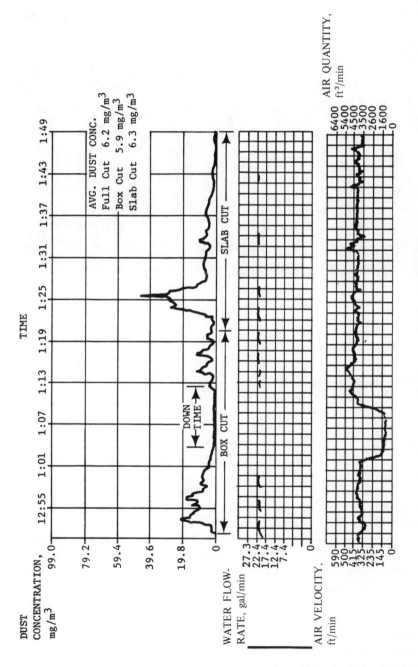

Figure 11.   Dust levels behind the brattice with the scrubber.

Figure 12.   Position of adjacent Simslins on double drum shearer face.

PROPOSED SAMPLING PLAN OF DUST CONTROL
TECHNIQUES FOR DOUBLE-DRUM SHEARERS

A major objective of this sampling program is to evaluate the effectiveness and reliability, under actual operating conditions, of respirable dust control strategies applicable to double-drum longwall shearers [BCR 1981c]. Nine underground coal mines having longwall operations will be sampled. Dust-control factors to be evaluated will be selected on a mine-by-mine basis. This selection will be based on the particular mine's operations and the operator's specific problems and interests.

A hypothetical sampling program might include such factors as cutting direction, water quantity, water pressure and airflow. Table IV defines a potential sampling plan for examining two levels of each of these four factors. The specific conditions of each test are listed in terms of the factors involved. The prescribed sequence of tests was determined so as to minimize the number of equipment changes and adjustments necessary to switch cutting directions, airflow, water pressure and water quantity. Specifically, the testing sequence is set up such that, during any one shift, only three equipment changes/adjustments are needed. These include one change in cutting

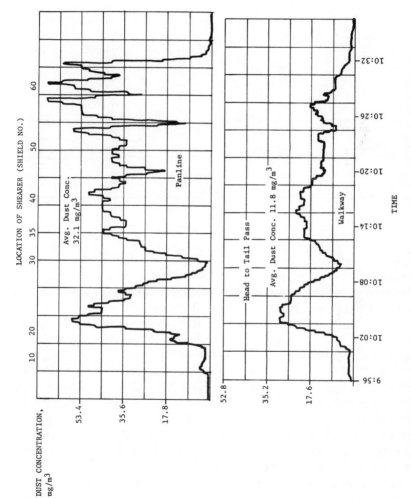

Figure 13.   Dust profiles obtained from adjacent Simslin samplers in a longwall face.

Table IV. Testing Sequence for Hypothetical Sampling Plan

Day	Test Number	Cutting Direction	Water Quantity (gal/min)	Water Pressure (psi)	Airflow (ft/min)
1	1	Head to Tail	50	300	250
	2	Tail to Head	50	300	250
	3	Tail to Head	50	150	250
	4	Tail to Head	70	300	250
2	5	Head to Tail	70	300	250
	6	Head to Tail	50	150	250
	7	Head to Tail	70	150	250
	8	Tail to Head	70	150	250
3	9	Tail to Head	70	150	400
	10	Head to Tail	70	150	400
	11	Head to Tail	50	150	400
	12	Head to Tail	70	300	400
4	13	Tail to Head	70	300	400
	14	Tail to Head	50	150	400
	15	Tail to Head	50	300	400
	16	Head to Tail	50	300	400

direction, one change in spray nozzles on drums, and one change in pressure adjustment at the pump. Occasionally, a change in cutting direction is needed between shifts in preparation for the next set of tests. Assuming that four passes can be completed per shift and that testing will take place during one shift per day, all 16 tests can be completed in four days. The fifth day can be used to make up for any unexpected delays or problems. As indicated in Table IV, the first two days of testing would occur with a 250-ft/min airflow level. After completing the first eight tests, airflow could be adjusted to 400 ft/min between the second and third day in preparation for tests 9 through 16.

In addition to being able to complete all 16 tests during the first week, there is another advantage to this particular testing sequence. After the first day of testing, preliminary indications of the effects associated with cutting direction, water quantity and water pressure can be obtained from the data. Specifically, a comparison of test conditions 1 and 2 would assess the effect of cutting direction. Comparing test conditions 2 and 3 would examine the effect of water pressure. Similarly, a comparison of test conditions 2 and 4 would indicate the effect of water quantity. Finally, a comparison of test conditions 3 and 4, along with information from the comparisons of 2 with 3 and 2 with 4, would provide an indication of the interactive effects of water pressure and quantity. In a similar fashion, each successive day of tests would provide additional information on the magnitude of the various individual factors and interactive effects on dust levels.

On completion of the first week of testing, a full analysis of the data will indicate if and under what set(s) of test conditions dust levels were sufficiently minimized. A second week of testing could consist of collecting additional data under the optimal conditions to allow increased confidence levels for dust level estimates obtained under these conditions. Alternatively, the second week of testing could consist of an exact replication of the first week's testing sequence, using a different drum. The drum could be changed easily over the weekend. The decision about the second week of testing must, obviously, be based on the priorities associated with testing a second drum vs building statistical confidence levels through replicating part or all of the previous testing.

It is important to note that the above, completely balanced, factorial design represents a hypothetical sampling plan for a given mine interested in these four control factors. Other sampling plans, having different factors and/or different numbers of factors, could be designed easily to suit the needs of a particular mine operation.

REFERENCES

BCR (1981a) "In-Mine Tests for Wetting Agent Effectiveness," Bituminous Coal Research, Inc., Final Report to BOM for Contract J0295091 (BCR Report L-1165).

BCR (1981b) "Machine Mounted Dust Collector," Bituminous Coal Research, Inc., BOM Contract No. H0377049, Prime Contractor Jeffrey Mining Machinery Company.

BCR (1981c) "Mine Demonstration of Longwall Dust-Control Techniques," Bituminous Coal Research, Inc., Research Program Proposal No. 258R for BOM RFP J0318001.

Cox, D. R., and E. J. Snell (1981) *Applied Statistics* (New York: Methuen, Inc.).

CHAPTER 20

USE OF LOGIT TRANSFORMATION TO ESTIMATE THE ERROR IN AEROSOL SAMPLING PARAMETERS

D. L. Stevens and O. R. Moss

Pacific Northwest Laboratory
Richland, Washington

ABSTRACT

The logistic distribution is a useful approximation for the normal distribution in aerosol sampling statistics. The concentration distributions are described by a mean and standard deviation, and the particle size distributions by one of the median diameters [e.g., mass median aerodynamic diameter (MMAD) and the geometric standard deviation (GSD)]. The precision of these estimated distribution parameters should be included as information useful for decision-making. This is normally not done because of the need for a large computer to do the regression analysis on the normal and log-normal distribution. The calculations would be made more often if they could be done on pocket, programmable calculators. We propose that the logistic distribution be used in these error calculations because its probability distribution function has a closed-form inverse. The regression analysis to estimate distribution parameters can be accomplished more concisely, leaving memory space in the calculator to perform confidence limit calculations.

THEORETICAL

Normal and log-normal distribution parameters are often determined graphically using probability paper. They can also be determined numerically using linear regression; however, this requires computation of the inverse normal cumulative distribution function. Although this computation can be done on a programmable calculator [Spencer and Lewis 1980], it is time-consuming and is likely to exhaust the memory capacity of the calculator. An alternative is to use the logistic distribution instead of the normal.

A standard form of the logistic cumulative probability distribution function is

$$F_X(x) = \{1 + \exp[-\pi(x - \mu)/(\sigma\sqrt{3})]\}^{-1} \qquad (1)$$

where μ = mean value
       σ = standard deviation of the random variable X

A useful property of the logistic distribution is that the logit transformation of $F_X(x)$ is a linear function of x:

$$\text{logit}(F_X(x)) = ln\{F_X(x)/[1 - F_X(x)]\} = \pi(x - \mu)/(\sigma\sqrt{3}) \qquad (2)$$

This property permits probability or proportion estimates restricted to the range [0,1] to be transformed into an unrestricted variable. The logistic cumulative distribution function has a shape similar to that of the normal cumulative distribution function:

$$G_X(x) = \int_{-\infty}^{X} \{\exp[-(s - \mu)^2/(2\sigma^2)]/\sqrt{2\pi\sigma^2}\} \, ds \qquad (3)$$

In fact, the maximum absolute value of the difference $|F_X(x) - G_X(x)|$ for $\mu = 0$, $\sigma = 1$ is less than 0.023. The maximum can be reduced to a value less than 0.01 by a change of scale in x in F_X using $F_X(16x/15)$ as an approximation to $G_X(x)$ [Johnson and Kotz 1970]. This degree of precision should be adequate for most applications. If greater precision is needed, Stevens and Moss [submitted] have developed a concise approximation to the normal distribution function based on a modified logistic distribution function that has a maximum absolute error of less than 3×10^{-5}.

Since $F_X(x)$ has an explicit inverse, it is feasible to perform the linear regression on programmable calculators and to include computation of precision estimates. Equation 2 has the general linear form

$$y = \beta_0 + \beta_1 x \qquad (4)$$

with

$$y = \text{logit}(F_X(x))$$
$$\beta_0 = -\pi\mu/(\sigma\sqrt{3})$$
$$\beta_1 = \pi/(\sigma\sqrt{3}) \tag{5}$$

The parameters β_0 and β_1 may be estimated using a linear regression routine. Usually the manufacturer of a programmable calculator will furnish a linear regression program. For completeness, we furnish the estimation equations in the appendix. The additional calculations necessary to obtain the variances and covariance of b_0 and b_1, the estimates of β_0 and β_1, respectively, are relatively minor and also given in the appendix. Equation 5 can be used to obtain estimates m of μ and s of σ from b_0 and b_1 as

$$m = -b_0/b_1$$
$$s = \pi/(b_1\sqrt{3}) \tag{6}$$

The usual form of a confidence interval as (estimate) ± (t value) × (standard deviation) is valid when the estimate has a normal distribution. In linear regression, the usual assumption of normally distributed errors implies that b_0 and b_1 have a bivariate normal distribution. It follows that m and s have a nonnormal distribution and the usual confidence interval formula should not be applied. Confidence limits on μ and σ can be obtained by applying Fieller's theorem [Finney 1964; Kendall and Stuart 1973]. The resulting α-level confidence limits are, for μ

$$\frac{m + h^2 v_{12} \pm h[h^2(v_{12}^2 - v_{11}v_{12}) + v_{11} + 2mv_{12} + 2m^2 v_{22}]^{1/2}}{1 - h^2 v_{22}} \tag{7}$$

and for σ

$$\frac{s(1 \pm h v_{22}^{1/2})}{1 - h^2 v_{22}} \tag{8}$$

where $v_{11} = \text{var}(b_0)$
$v_{22} = \text{var}(b_1)$
$v_{12} = \text{cov}(b_0, b_1)$
$h = t_\alpha/b_1$

and t_α is the α-level student's t variate with $N - 2$ degrees of freedom, where N is the number of data points.

EXAMPLE APPLICATION

The technique is illustrated for the simple approximation of the normal cumulative distribution function, $G_X(x)$, with a logistic distribution, $F_X(16x/15)$ [Johnson and Kotz 1970]. The results are compared to an analysis using a normal distribution. The data used were from a sodium chloride aerosol from Pacific Northwest Laboratory. The data are given in Table I. The analyses were carried out assuming that particle diameters followed a log-normal distribution so that the x in Equation 4 was taken in ln(diameter) and the y was calculated to be either $G_X^{-1}[ln(\text{diameter})]$ or $F_X^{-1}[16(ln(\text{diameter}))/15]$. The results may be compared in Table II.

The estimates obtained using the different distributions are quite similar. The log-normal distribution gives a smaller confidence region; however, the log-logistic estimates are within the log-normal confidence and conversely.

Table I. Example Data: NaCl Aerosol

D, Cutoff Diameter (μm)	F, Cumulative Weight Fraction Smaller Than D
0.29	0.005
0.51	0.036
0.79	0.124
1.70	0.364
2.67	0.584
3.92	0.857
5.90	0.941
8.11	0.971

Table II. Distribution Parameter Estimates and 95% Confidence Limits

Parameter	Log-Logistic Estimate of Log-Normal			Log-Normal		
		95% Limits			95% Limits	
	Estimate	Min.	Max.	Estimate	Min.	Max.
MMAD	2.03	1.87	2.20	1.97	1.85	2.11
GSD	2.04	1.94	2.16	2.10	2.01	2.20
R^2	0.991			0.995		

APPENDIX

The formulas for estimating the parameters β_0 and β_1 of Equation 4 are given below, with b_0 the estimate of β_0, and b_1 the estimate of β_1:

$$b_1 = \frac{\Sigma x_i y_i - (\Sigma x_i)(\Sigma y_i)/N}{\Sigma x_i^2 - (\Sigma x_i)^2/N}$$

$$b_0 = (\Sigma y_i - b_1 \Sigma x_i)/N \tag{A1}$$

The quantities that need to be accumulated and stored to calculate b_0 and b_1 are Σx_i, Σx_i^2, Σy_i, and $\Sigma x_i y_i$. If, in addition, the quantity Σy_i^2 is saved, then an estimate of the variance about the regression is given by

$$s^2 = b_1 [\Sigma x_i y_i - (\Sigma x_i)(\Sigma y_i)/N] - [\Sigma y_i^2 - (\Sigma y_i)^2/N] \tag{A2}$$

This estimate has $N - 2$ degrees of freedom.

From the same quantities one can obtain the variances and covariance of b_0 and b_1 [Draper and Smith 1966]:

$$var(b_0) = v_{11} = \frac{s^2 \Sigma x_i^2}{N \Sigma x_i^2 - (\Sigma x_i)^2}$$

$$var(b_1) = v_{22} = \frac{s^2}{\Sigma x_i^2 - (\Sigma x_i)^2/N}$$

$$cov(b_0, b_1) = v_{12} = \frac{-s^2 \Sigma x_i}{N \Sigma x_i^2 - (\Sigma x_i)^2} \tag{A3}$$

Although Equation A3 can be used to construct confidence limits on b_0 and b_1, we want confidence limits on μ and σ. These can be obtained by applying Fieller's theorem: if a and b are unbiased estimates of α and β with a multivariate normal distribution, and if the estimated variances and covariance, based on N degrees of freedom, of a and b are s_{11}, s_{22} and s_{12}, respectively, then the α-level confidence limits for the ratio α/β are the roots of

$$x^2(b^2 - t^2 s_{22}) - 2x(ab - t^2 s_{12}) + (a^2 - t^2 s_{11}) = 0$$

where t is the appropriate deviate with N degrees of freedom at the α probability level.

To get confidence limits for μ, we apply Fieller's theorem with a = $-b_0$, and b = b_1. The solution of the quadratic equation in the theorem is given by Equation 7. To apply the theorem to σ, take a = $\pi/\sqrt{3}$, b = b_1. Then var(a) = cov(a,b) = 0, and the solution of the quadratic is given by Equation 8.

ACKNOWLEDGMENTS

This work was supported by the U.S. Department of Energy, under contract DE-AC06-76RLO 1830.

REFERENCES

Draper, N. R., and H. Smith (1966) *Applied Regression Analysis* (New York: John Wiley & Sons, Inc.).

Finney, D. J. (1964) *Probit Analysis* (Cambridge: Cambridge University Press), p. 63.

Johnson, N. L., and S. Kotz (1970) *Distributions in Statistics: Continuous Univariate Distributions, Vol. 2* (New York: John Wiley & Sons, Inc.).

Kendall, M. G., and A. Stuart (1973) *The Advanced Theory of Statistics, Vol. 2* (New York: Hafner).

Spencer, B. B., and B. E. Lewis (1980) "HP-67/97 and TI-59 Programs to Fit the Normal and Log-Normal Distribution Functions by Linear Regression," *Powder Technol.* 27:219.

Stevens, D. L., and O. R. Moss (submitted) "Simple Approximations to the Normal and Inverse Normal Distribution Functions," *Am. Statistician.*

CHAPTER 21

APPLICATION OF A STRATIFIED RANDOM SAMPLING TECHNIQUE TO ESTIMATION AND MINIMIZATION OF RESPIRABLE QUARTZ EXPOSURE TO UNDERGROUND MINERS

C. E. Makepeace

Ottawa, Ontario

F. J. Horvath and H. Stocker

Atomic Energy Control Board
Ottawa, Ontario

ABSTRACT

The aim of a stratified random sampling plan is to provide the best estimate (in the absence of full-shift personal gravimetric sampling) of personal exposure to respirable quartz among underground miners. One also gains information of the exposure distribution of all miners at the same time. Three variables (or strata) are considered in the present scheme: locations, occupations and times of sampling. Random sampling within each stratum ensures that each location, occupation and time of sampling has equal opportunity of being selected without bias. Following implementation of the plan and analysis of collected data, one can determine the individual exposures, the individuals or groups having the highest exposures, the distribution of exposures, and the mean. This information can then be used to identify those groups whose exposure contributes significantly to the collective exposure. In turn, this identification, along with other considerations, allows the mine operator to carry out a cost-benefit optimization and eventual implementation of engineering controls for these groups. This optimization and engineering control procedure, together with the random sampling plan, can then be used in an iterative manner to minimize the mean value of the distribution and collective exposure.

INTRODUCTION

Exposure to respirable quartz has long been recognized as an occupational hazard. The diseases resulting from these exposures are of the pneumoconiosis type, of which silicosis is one example. These diseases appear in the lung and, generally, are chronic or slowly developing. In Canada, silicosis was observed in the early years of underground mining among metal and nonmetal miners, generally where poor ventilation conditions existed or where dust suppression techniques were absent. As health authorities became aware of the relation between silicosis and the airborne hazards that may have contributed to its occurrence, the agencies with authority over mining took steps to have the mining companies introduce forced ventilation and dust suppression techniques to reduce the concentrations of airborne hazards, including respirable quartz.

In underground mines in which quartz is present in the host rock, it is widely recognized that personal exposures to the respirable fraction of quartz must be determined and monitored accurately. Not only is this true for regulatory and compliance purposes, where the health protection of individuals must be afforded, but also there is the desirability to minimize, by suitable engineering controls, the collective exposure to all the miners. As used in this chapter, the attributed collective exposure (or the consequent biological collective dose) means the sum of all exposures assigned to all workers in a group over any given period of time. The principle of minimization of the collective exposure of the group is applied below the relevant exposure limit for individuals. It recognizes the possibilities of individual susceptibilities to the specific toxic substance and aims at minimizing the risk of the toxicant to the subgroups having the highest collective exposure.

It is also well known that personal, full-shift, gravimetric sampling is the best means of estimating individual exposures to respirable quartz. However, in the absence of a system of personal full-shift gravimetric sampling, it may be necessary to base exposure estimation for individuals on a system of grab sampling (using konimetric methods) combined with time weighting, that is, a sum of products of the particle number concentrations and the times spent in these concentrations. It is recognized that the particle counts determined by the konimeter still must be related to the mass of respirable quartz. Although it is obvious that this latter system has its limitations, including that of associating particle number concentrations with respirable quartz concentrations, the limitations associated with a nonrandom sampling scheme can be reduced substantially by using a scheme that is both random and stratified. Random sampling ensures that each location, occupation and time of sampling has equal opportunity of being selected without bias. In this case, stratification means that each variable (or stratum), such as location, occupation and time, can be ordered or arranged such that each element (such as specific location, occupation or time interval) can be selected randomly, in turn, using the techniques of probability theory. The stratified sampling plan can apply to full-period single samples (a personal sampler distributed to a different miner each day), full-period consecutive samples (each location sampled in each time interval) or grab samples (a random selection of samples at various time intervals). The application of a grab sample plan is described in this chapter. It should be noted in passing that the stratified random sampling technique can be used equally for dust, radon daughters, or other similar industrial or mining aerosols.

THEORETICAL

Establishment of Baseline Information

Baseline information must be obtained for each location (and occupation, if needed). This may be done by a series of successive grab samples taken in a given location, thus establishing a mean and standard deviation for this location. (To take account of the inherent variability of concentrations of respirable quartz in underground mining environments, this procedure may have to be done several times, perhaps at different times of the day, or on different days.) For each such set of measurements, one calculates the minimum number of measurements required to establish an individual's exposure in that location, say with an uncertainty of 50% at the 95% confidence level [ANSI 1973]. When all such locations have been sampled in this way, a sampling time interval can be assigned conveniently to accommodate all such locations. If any location has a requirement for sampling not provided for in the scheme above, the appropriate engineering controls (such as ventilation or dust suppression) can be increased so that concentrations fall to levels reasonably consistent with the other locations in the sampling scheme. (If sufficient controls cannot be imposed on any given location, that location must be placed in a subset of locations to be sampled separately.) At this stage, a "steady-state" condition has been achieved in this set of locations and the stratified random sampling plan can be implemented.

If an integrating system is used to determine the initial average baseline concentration, grab sampling must still be done to establish the minimum sampling frequency for a given location or occupation. An example of the use of gravimetric dust sampling is given in the following. Nine occupations were studied by the authors in one set of observations in an underground uranium mine in Canada. Each occupation has been given an identifying number for convenience of reference, for example, scooptram operator (38), conveyor belt operator (47) and so on. Measurements of respirable quartz have been made for the nine occupations (Figure 1). In Figure 1, it is clear that the concentrations of respirable quartz associated with the conveyor belt operator are much greater than those for the other occupations studied. Therefore, that occupation should be considered as part of a set different from the subset which could be formed by the other eight occupations. This information can then be used as input to the baseline described above.

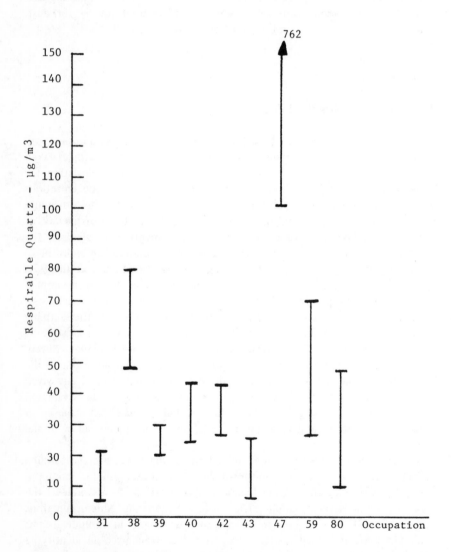

Figure 1.   Personal dust sampling observations (for respirable quartz) in uranium mines for various occupations.

Description of Stratified Random Sampling Plan

Precise predictions of the concentrations of respirable quartz in dust concentrations depend on the assumption that mine ventilation will not be interrupted except for short periods. Unusual occurrences, such as fan failures, generally cannot be predicted. Stratified random sampling procedures will ensure that each of these unusual occurrences will have equal probability of being included in an unbiased location-time sampling scheme.

The example presented below is designed to describe methods of obtaining simplified random samples of location-occupation-time for underground workers in a mine. Since these are finite-sampled populations or populations of known size, the number of items in the population from which the sample is to be selected is always known. Selecting a random sample from a finite population will ensure that each possible sample and each item in the population has an equal opportunity of being included in the sample selected. Because of the large number of mine locations involved and the variation in the operable duration of each of these locations, time and motion studies, together with the application of statistical design of stratified random sampling techniques, will ensure a more efficient use of sampling man-hours.

As an example of the stratification of shift sampling times [Makepeace 1981], suppose one wishes to select a total of six random samples for an eight-hour shift period and an individual sampling time of 10 minutes. The total number of sampling intervals n will have a value of n = 6 sampling intervals/hr X 8 hr = 48 consecutive sampling intervals per eight-hour shift (Table I). One then requires access to a random number table such as that

Table I. Ten-Minute Sampling Intervals and Their Corresponding Times of Day

Sampling Interval	Time of Day
1	8:00–8:10 AM
2	8:10–8:20 AM
3	8:20–8:30 AM
4	8:30–8:40 AM
5	8:40–8:50 AM
.	.
.	.
.	.
Lunch–12:00–12:30 PM	
.	.
.	.
.	.
46	3:30–3:40 PM
47	3 40–3:50 PM
48	3:50–4:00 PM

given in Table II [Natrella 1963]. Entry to the table may be based on some subjective choice, such as the last two digits of one's social insurance number, social security number, telephone number or otherwise. Two-digit numbers (less than 48) are selected, since we are dealing with 48 sampling intervals. Suppose the last two digits of someone's telephone number are 16. Entry to the table is made at number 16, and 5 additional numbers (reading down the columns) are selected, as shown. The 48 ten-minute consecutive sampling intervals for the eight-hour shift are arranged into a judgment grab sampling chart (Table III), in which the six selected random samples are again indicated. These six random samples now correspond to specific and sequential ten-minute intervals in the eight-hour shift (Table IV). Repeated use of the same portion of any random number table is invalid and does not satisfy the intent of randomness when selecting a sample. Instructions for the use of various available tables of random numbers, including repeated access to them, have been prepared [Makepeace 1981].

Table II. Typical Random Number Table (Two-digit)

03	–18–	68
97	– 9–	74
–16–	–23–	27
–12–	52	00
55	37	29
16	70	16
84	56	11
63	99	35
–33–	16	38
57	37	31

Table III. Judgment Grab Sampling Chart—Ten-Minute Periods

Consecutive Shift Time (hr)	Minutes					
	10	20	30	40	50	60
1	1	2	3	4	5	6
2	7	8	– 9–	10	–11–	12
3	13	14	15	–16–	17	–18–
4	19	20	21	22	–23–	24
5	25	26	27	28	29	30
6	31	32	–33–	34	35	36
7	37	38	39	40	41	42
8	43	44	45	46	47	48

Table IV. Six Randomly Selected Sampling Intervals and Their Corresponding Sequential Time Intervals for an Eight-Hour Shift

Sampling Interval	Time of Intervals
9	9:20– 9:30 AM
12	9:50–10:00 AM
16	10:30–10.40 AM
18	10:50–11:00 AM
23	11.30–11:40 AM
33	1:20– 1:30 PM

Confidence Limits

There is an inherent variability in the workplace concentrations of respirable quartz. A limited number of observations of concentrations may not be representative of the population mean value or of the entire range of possible values. The evidence of many experimental trials as well as rigorous mathematical proof shows that it is possible to make inferences relating the mean or average of a small sample ($n \leqslant 30$) to the grand average that would be obtained if a great many more small samples were obtained. It should be realized that a good small sample is better than a poor large sample, and that a quite moderate amount of data submitted to efficient statistical evaluation often makes a much more convincing case.

When one calculates the confidence limits for the mean from the data obtained from a small sample, what one is actually doing is creating a range (minimum and maximum) within which the population mean will lie, say, 99, 95 or 90% of the time. This mean is not the mean of the particular small sample, but the average of all the means one would obtain if, instead of just one sample, one had hundreds of samples obtained in like manner. For instance, suppose one has obtained 30 values from a particular location. The confidence limits one calculates at the 95% confidence level will give a minimum and a maximum that will contain the average of hundreds of such means of observations, should one have continued to obtain these averages over a period of months, provided, of course, one obtained the sets of data under like circumstances. That is all the confidence limit tells us, and one should bear in mind that 5 means out of 100 will lie outside this range at the 95% confidence level. If one calculated the 99.999% confidence interval, one mean in 100,000 would lie outside the range. However, this range will be very much broader or wider.

If one adopts a 95% level of confidence, the statement one can make about the average of the population of individual measurements is as follows: the probability is approximately 0.95 that the average of the population of individual measurements is included between the 95% confidence limits:

$$X \pm t_{0.975} \frac{s}{\sqrt{n}}$$

t is the student t value, and $t_{0.975}$ depends on n, the sample size, and the number of degrees of freedom (df) = n − 1; s is the sample standard deviation. (The correct t value column for the calculation of the 95% confidence level is $t_{0.975}$).

The correct statistical interpretation of these 95% confidence limits is the following: suppose many random samples of size n are drawn, and that the confidence limits

$$X \pm t_{0.975} \frac{s}{\sqrt{n}}$$

are calculated for each sample. In about 95% of the samples, it will be found that the confidence limits include the average of the population between them. The precise form of this statement should be noted very carefully. It is widely misunderstood and misused. In particular, it should be observed that the statement does not mean that about 95% of the individual measurements in the population will be included between the limits

$$X \pm t_{0.975} \frac{s}{\sqrt{n}}$$

yet this is the interpretation often given to it.

The solution to the problem of dealing with the inherent variability in the workplace concentrations of respirable quartz is the use of a stratified random sampling technique followed by the calculation of the confidence limits for the universe or population mean at, say, the 95% probability level.

Implementation of Sampling Plan

Grab sampling is carried out according to the plan shown in Table III. The concentrations of respirable quartz measured in each location are multiplied by the time each worker spends in that location. The sum of all such products for each individual is the "best estimate" of exposure for that individual under this particular sampling plan.

From the individual "best estimate" of exposure for each worker, a histogram may be constructed of the worker population studied. In turn, workers' exposures may be grouped by occupation or location to reveal occupations or locations having the highest exposures associated with them. Since exposure control is the ultimate aim of the engineering and monitoring program, engineering controls (such as improved ventilation and dust suppression techniques) can be imposed to reduce the concentrations so measured. Moreover, if several control measures are required, the mine operator can address these one at a time, in the most advantageous cost/benefit manner. The cost of the monitoring program should be included in any cost/benefit considerations. Measurement of the exposures following the imposition of a specific engineering control indicates the effectiveness of the control technique. If additional engineering controls are required, they can be imposed in an iterative manner.

CONCLUSIONS

A method for optimizing the sampling time available has been introduced using a stratified random grab sampling plan. This plan gives each sampling interval, location and occupation equal opportunity of being selected in an unbiased manner. Combined with some method of time accounting, it gives the "best estimate" of personal exposure estimation based on a grab sampling scheme.

A histogram of the personal exposures for all workers allows the collective exposure to be calculated, and it serves to reveal other quantitative features, such as the mean value of the distribution of exposures and the number of persons having exposures above any arbitrary value. It also serves to highlight the subgroups having the highest collective exposures. Such information on the individual exposures, collective exposures, locations, and amount and cost of ventilation and other control techniques, can then be used in a cost/benefit optimization for the subgroup having the highest exposures and associated risks. Should the reduction in collective exposure outweigh the costs of implementing those engineering controls needed to produce such reduction, the operators of such mines should carry out the necessary changes. In this way, the collective exposure for the most highly exposed subgroup is reduced (as is that of the entire group) and the mean exposure for the group is reduced as well (in all likelihood). In an iterative manner, the random sampling plan is then reintroduced and a new baseline is established, perhaps with different subgroups. Further identification of new high-exposure subgroups followed by cost/benefit optimization and further engineering controls continue the iterative procedure and reduction of collective exposure.

REFERENCES

ANSI (1973) "Radiation Protection in Uranium Mines," American National Standards Institute, ANSI-N13.8, New York.

Makepeace, C. E. (1981) "Stratified Random Sampling Plans Designed to Assist in the Determination of Radon and Radon Daughter Concentrations in Underground Uranium Mine Atmospheres," Atomic Energy Control Board, INFO-0038, Ottawa, Ontario.

Natrella, M. G. (1963) "Experimental Statistics," National Bureau of Standards Handbook 91, U.S. Government Printing Office, Washington, DC.

CHAPTER 22

MATHEMATICAL BASIS FOR AN EFFICIENT SAMPLING STRATEGY

Nurtan A. Esmen

Department of Industrial Environmental Health Sciences
Graduate School of Public Health
University of Pittsburgh
Pittsburgh, Pennsylvania

ABSTRACT

The Corn-Esmen sampling strategy based on workplace exposure zones is an efficient way to obtain workplace environmental quality data. In this chapter, the mathematical extension of this strategy is developed to include the use of the data as a basis for making decisions on the necessity of preventive measures to protect worker health. The results suggest the necessity of sampling a minimum of five to seven workers for two to three appropriate sampling periods. The method of data analysis proposed suggests the calculation of the confidence level of the appropriate conclusions based on the data rather than the calculation of numerical bounds for the results.

INTRODUCTION

Sampling is a methodical information-gathering process; it is rarely more than a tool that serves a distinct, presumably well defined purpose. The purpose of sampling determines the nature of sampling strategy as well as the methodologies of sample collection and analysis. Although the theoretical development presented here is sufficiently general to extend to cover a wide variety of circumstances, the discussion is limited to environmental surveillance for risk assessment purposes. Furthermore, it is stressed that the thrust of this chapter is for nonresearch purposes. There is a crucial distinction between research and nonresearch health hazard assessment; this distinction was pointed out by Drinker and Hatch [1936]:

> The aim of dust sampling and determination is not to measure the absolute dust concentration in the air but rather to obtain an index of the health hazard involved in breathing the dust-laden air. Requirements are determined by physiological considerations and not by the physical criteria usually applied to precise instruments.

This statement applies fully to any chemical agent (dusts, gases and vapors) which may be involved in hazard evaluation. A number of different situations also exist in environmental surveillance. Sampling as it pertains to enforcement of legal standards entails a number of requirements that usually are determined through a consensus rather than a scientific process. Although the aim of environmental surveillance for enforcement purposes does not differ from the aim of environmental surveillance for risk assessment purposes, the legal restrictions do not necessarily allow the flexibility necessary for professional judgment pertaining to the hazard control.

Sampling as related to risk assessment in a workplace is inherently the assessment of the probability and/or frequency of occurrence of an undesirable event. In this perspective, it is solely a vehicle to quantify an existing condition so that a basis for professional judgment is established in regard to the necessity and/or extent of preventive measures to be taken. Obviously, this judgment does not explicitly depend on the results of sampling, but must include other inputs, such as the toxicological behavior of the agent.

In any generalized or specific sampling scheme to fulfill the needs of environmental surveillance of a workplace under a variety of circumstances, the objectives of sampling must be clearly defined and realistic. The highly complex nature of the workplace can easily lead to a situation where the objectives of sampling become secondary to the details of methodologies, sampling strategies, statistical treatment of data resulting in unrealistic expectations or impossible demands placed on the conclusions that can be drawn from such a survey. In addition, a sampling scheme must be efficient: the information necessary to make a decision should not require an enormous number of samples. These considerations lead to the necessity of careful planning of a sampling strategy.

Several steps can be taken at the planning stage to maximize the information obtained by a minimum number of samples. One strategic device that may be used is the stratification of the sampled population. A stratified sampling strategy based on the classification of the workers in so called "exposure zones" was proposed by Corn and Esmen [1979]. The exposure zones can be, but are not necessarily, a definable area of the physical plant.

By definition, the zones are based on the similarity of agent, work and environment for the workers classified in a zone, and thus contain a distinct relationship to the worker exposure characteristics. Since the "zoning" is based on an agent, as well as work practice, a worker may belong to different zones for different agents. The statistical treatment of observations in this strategy is based on the log-normality of the measured airborne concentration distributions [Esmen and Hammad 1977], and for any selected zone the limiting probability of measured exposure concentrations C to be greater than a prespecified concentration C_r for the agent is expressed in terms of an exposure parameter ϕ

$$\phi = \frac{ln\,(C_r/M)}{ln\,S} \qquad (1)$$

where C_r = applicable criterion
        M = geometric mean of measured concentration distribution
        S = geometric standard deviation of the measured concentration distribution

In mathematical terms, this limiting probability is expressed as:

$$\Pr(C > C_r) = \frac{1}{2}\,[1 - \mathrm{erf}(\phi/\sqrt{2})] \qquad (2)$$

The exposure parameter ϕ defines the excess (or deficit) in a manner free from the value of criterion; more importantly, this parameter can be used as a basis for decision-making.

THEORETICAL CONSIDERATIONS

To develop the model, we consider a zone of workers from which W workers are randomly selected and sampled for D number of periods. The duration of the sampling period is related to the agent sampled; the appropriate duration for an agent may be determined from the applicable considerations suggested by Roach [1966, 1977]. For simplicity we shall assume eight-hour samples only and consider D to be the number of days. It is important to note that this sampling scheme and for that matter, any environmental sampling scheme that involves repeated measure of exposures, does not constitute WD independent observations. Therefore, the calculation of the parameters involved must include a proper statistical mode. If we index the measured concentrations among workers by i and between days by j, and express the model on the basis of log-normality of the distribution of the concentration, then we expect the logarithmic concentration to be distributed about a mean with an internal normal distribution, i.e., $Y_i \sim N(0, \omega^2)$ and an "error" term, $Z_{ij} \sim N(0, \sigma^2)$. Naturally, this model assumes the measurements to be unbiased, and hence:

$$ln\, C_{ij} = \overline{ln\, C} + Y_i + Z_{ij} \qquad (3)$$

It should be noted that this model is the estimation of the experience of sampled workers and thereby the estimation of the experience of the zone population over extended periods of time. The sampling is done looking at time segments of the events; therefore, the model carries the inherent assumption that over the period in which the estimates are going to be used, the long-term trend of exposures do not have a significant contribution. That is, over the period in which the estimates will be used, there is no specific increase or reduction of the exposure. With this constraint, the error term includes the variability of both the process and sampling error.

To develop an understanding of the behavior of the exposure parameter, let

$$\psi = ln\, C_r - \overline{ln\, C} \qquad (4)$$

Thus, for WD number of samples with θ^2 as the variance of $ln\, C_{ij}$, we get:

$$\frac{\psi}{\sqrt{\theta^2}} \sim N(0, 1) \qquad (5)$$

Consequently, the estimate of $ln\, C_r - \overline{ln\, C}$, $\hat{\psi}$, may be expressed as:

$$\frac{\hat{\psi}}{\theta} - \phi = N(0, 1)/\sqrt{WD} \qquad (6)$$

By replacing θ with its estimate based on sample results, and rearranging the equation in terms of estimate of ϕ, we get:

$$\hat{\phi}\sqrt{\frac{X_\alpha^2(WD - 1)}{WD}} + \frac{U_\alpha}{\sqrt{WD}} < \phi < \hat{\phi}\sqrt{\frac{X_{1-\alpha}^2(WD - 1)}{WD}} + \frac{U_{1-\alpha}}{\sqrt{WD}} \qquad (7)$$

U_α is the unit normal variate at α level. Equation 7 describes the appropriate α confidence bounds of ϕ for W workers sampled D days with $\hat{\phi}$ estimate of $\hat{\phi}$ obtained from the data.

This interval with p = 1 – 2α level of confidence is an important concept in the development of the proposed sampling strategy and, furthermore, in understanding the limitations inherently placed on environmental surveillance by our limited ability to obtain large number of samples. It is important to note that increasing sample size by shortening the sampling period does not necessarily improve the model. If the appropriate sampling period as determined by the closest practical sample duration near the Roach criteria is replaced by a number of samples of shorter duration, another term will have to be included in the model (Equation 3). Generally, this inclusion is to the detriment of the efficiency of the estimation process, because, by sampling over the entire (or nearly entire) appropriate period, we are in fact pooling a very large number of samples in the determination of a mean value for that period.

As a justification for a new strategy, let us first use the bounds determined by Equation 7 in an example to show the limitation of the concept of using environmental surveillance data as an absolute measure. Suppose that, due to a legal or administrative restriction, we do not wish to exceed a given level of concentration at all. Suppose that previous experience suggests that ϕ is within 90% of a critical value for the undesirable event, and if we have 90% confidence in this value we can claim a satisfactory achievement of the objective of not exceeding the specified concentration. In other words, based on the variability of the analytical method and process if we can show that the chance of "not getting caught" is sufficiently good, then we claim success. In reality, we may very rarely, if at all, experience a true excursion in exposure beyond this critical level. Consequently, if the objective of the sampling becomes to "prove" or "disprove" this condition, the statistical objective is to construct a 90% confidence interval of 90% tolerance level of the estimate. If the zone where such an estimate is to be made contains, for example, 20 workers, Equation 7 may be evaluated for the necessary number of samples. The number for this example turns out to be >5300 total samples, or sampling the entire population for >265 working periods. Such a requirement would clearly be absurd. Thus, the decision-maker will be left with no choice but to recommend the installation of control devices to lower ϕ to such an extent that the demonstration of $\hat{\phi} < \phi_{Cr}$ is possible with a reasonably few samples. Such a recommendation may conceivably be at the expense of an existing serious problem at a different zone in which a surveillance is yet to be carried out; furthermore, the preventive measure may be likely to carry no significant impact on improved healthfulness of the work place.

The philosophical basis of sampling for risk assessment purposes is to gather information on the necessity and/or degree of preventive measures to be undertaken. On this basis, the objective of sampling can be worded in a slightly different manner: "suppose a fraction of workers in a zone are sampled for a few days and an estimate of exposure parameter ϕ is obtained. With this information, what can be said about the exposure conditions in the workplace?" This objective is directly related to the desired outcome; furthermore, sampling is used in a manner designed to enhance the quality of the outcome (decision) rather than the quality of an intermediate step (sampling results).

By the stratification process, the selected zone represents the daily experience of N workers, and these experiences span an intermediate number of periods d. For any prespecified d periods we have an event space n such that Nd events take place. If these events, that is the values of exposure concentration, are ordered, then the probability of Kth occurrence of the undesirable event $C_K > C_{crit}$ may be calculated in a straightforward manner:

$$\Pr(X = k) = \Pr(C_k > C_{crit}) = \sum_{w=k}^{n} \frac{n!}{w!(n-w)!} [\Pr(-\phi)]^{n-w} \cdot [1 - \Pr(-\phi)]^{w} \qquad (8)$$

Equation 8 can be used to calculate the probability of occurrence once, twice, etc., by setting $x = 1, 2, \ldots, n$. However, the probabilities of interest are generally for one occurrence $(X = 1)$ and the probability of the desirable event where no over exposure occurs $(X = 0)$:

$$\Pr(X = 0) = \Pr[\max(C) < C_{crit}] = [\Pr(\phi)]^{n} \qquad (9)$$

Here we note that for all values of $\phi < \infty$, there is going to be a distinct probability (however small) of $\max(C) > C_r$ and, as n increases, this probability will also increase. In other words, if we sample all workers long enough, we are bound to show an excess. In terms of unhealthful consequences, the magnitude and frequency of occurrence of this excess is the only meaningful measure. By setting the critical level C_r sufficiently low, a certain amount of control on the magnitude of this excess can be achieved. If the excess over a sufficiently low setting of the critical level is a rare event, then a very large excess over the true critical level would be an even rarer event. The significance of the frequency of excess is a judgmental proposition. The proper considerations lie in the consequences of overexposure at infrequent intervals, level of protection inherent in the setting of the critical level, the magnitude of the consequences of even a single overexposure, and the accuracy of the analytical technique. For example, if a single overexposure can possibly result in death or serious health damage, it would be foolhardy to tolerate such an event even at a frequency of once every ten years, even if the analytical technique or process changes do not permit a very refined measure. In contrast, if the consequences of a single overexposure are within the homeostatic capabilities of the human body or if it is possible transient irritation with no significant health consequences, then a distinctly significant probability of single occurrence in a shorter period (year, month or even a week) may be tolerated. These considerations involve the judgment of safety in contrast to evaluation of risk.

As a simple tool but not as a strategic device, Equations 8 and 9 may be used to determine the probabilities associated with zero and one or more occurrences. To aid this process, these probabilities are graphically shown as a function of number of worker days and the estimated exposure parameter in Figures 1 and 2. The evaluation is a straightforward process, the estimated exposure parameter assigns probability of zero occurrences and one or more occurrences for the entire group of workers over a number of days. For example, if the estimate of ϕ from a sampling exercise were found to be 3, and if there were 25 workers in that zone, then the probability of zero overexposure in a workweek would be nearly 90%, and probability of one or more overexposures would be less than 10%. However, this straightforward procedure does not address the questions of decision-making, sampling

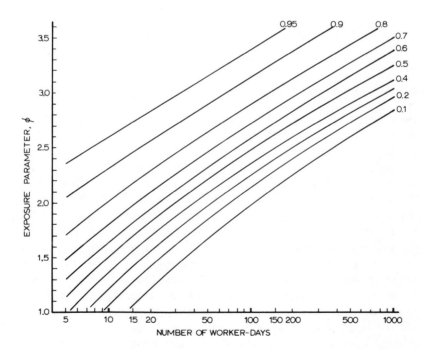

Figure 1. Zero overexposure probabilities shown as a function of exposure parameter ϕ and number of worker-days.

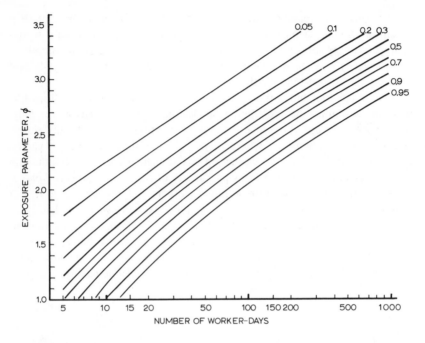

Figure 2.   One overexposure probability shown as a function of exposure parameter ϕ and number of worker-days.

strategy and confidence levels. Therefore, these questions still must be addressed to develop a certain process in which the results can be used to aid decision-making.

Suppose we choose a priori levels of acceptable frequency of overexposure based on the available knowledge on the agent. Then we can make the claim that if the probability of zero overexposure is larger than the critical probability of zero exposure (a priori decision) at the corresponding $n_0 = Nd_0$ value, then we make the claim of no overexposure; similarly, if the probability of one overexposure is larger than the critical probability of one overexposure (a priori decision) at the corresponding $n_1 = Nd_1$ value, we make the claim of overexposure. To each of these assigned event spaces and probabilities there exist unique values of ϕ_0 and ϕ_1, respectively. If the estimate of ϕ from sampling is $> \phi_0$ or $< \phi_1$, we make the proper decision and calculate the confidence we have in this decision. This procedure will result in three possibilities: $\hat{\phi} > \phi_0$, $\hat{\phi} < \phi_1$ and $\phi < \hat{\phi} < \phi_1$. In the first and second cases there is a clear-cut decision, but in the third case neither decision can be made directly, but we can still select the best available decision.

If the decision indicates no overexposure, we would like to see that the lower confidence limit of ϕ is greater than ϕ_1. In fact, we can use this calculation as the confidence level of decision by simply calculating the value of the lower confidence interval which corresponds to ϕ_1. Conversely, if the decision indicates overexposure, we calculate the α value of the upper confidence interval which corresponds to ϕ_0. In the third condition, we calculate the confidence level of each decision and compare the respective α values.

Clearly, the confidence level of each decision is based on the relative position of ϕ_x with respect to $\hat{\phi}$ as a function of the sample size. We would like to keep the confidence interval thus calculated as short as possible, so that the confidence on the decision would be as high as possible. If we analyze the worst case, $(\phi_1 < \hat{\phi} < \phi_0)$ as the limiting criterion and call this interval arbitrarily the indecision interval, then it can be claimed that the "best" decision would be over the maximum length of this interval or when $\hat{\phi} = \phi_1$ or $\hat{\phi} = \phi_0$. The length of this interval is determined a priori; therefore, the shorter the confidence interval about ϕ the higher the confidence level on the decision we can have. In this region the normalized length of the interval may be estimated by:

$$\frac{\phi_1 - \phi_0}{\hat{\phi}U_{1-\alpha}} = \frac{2}{\sqrt{WD}} \qquad (10)$$

The confidence level of the decision, although slightly better, does still vary nearly inversely proportionally to the square root of the number of total samples taken. This observation indicates that beyond the value of $WD \sim 40$ we gain but little and below $WD \sim 10$ the maximum interval would be too large to make a clear decision sufficiently often. Therefore, as the first step of a strategy, a minimum acceptable number of samples may be proposed to be 5–7 workers sampled for 2–3 days. If there are fewer than 5 workers in a zone, it would be necessary to sample all workers a sufficient number of times to get $WD \sim 12$.

The basis of such a sampling strategy does not deviate from the strategy of stratification proposed by Corn and Esmen [1979]. In the zones selected, the sample size may be set by the considerations discussed above. However, interpretation of the results needs to be carried out in an entirely different manner. In the interpretation process we select a basis for decision-making and from the data we proceed to calculate the confidence we have in the decision. Suppose from toxicological considerations, a level of ϕ_1 is selected to be the bound for overexposure and ϕ_0 is selected to be the bound for zero exposure. From the data obtained we can calculate the lower bound of ϕ by Equation 7 at any confidence level desired. Consequently, if we set ϕ to be greater than ϕ_1 (negation of one or more overexposure) this negation will be achieved at a confidence level $p = 1 - \alpha$ when the corresponding confidence limit is ϕ_1. Therefore,

$$\phi_1 \left[\frac{X^2(WD - 1)}{WD} \right]^{1/2} + \frac{U_{1-\alpha}}{\sqrt{WD}} = \hat{\phi} \tag{11}$$

The term containing X^2 may be estimated in terms of $U_{1-\alpha}$ as:

$$\left[\frac{X^2(WD - 1)}{WD} \right]^{1/2} \doteq \frac{U_{1-\alpha}}{\sqrt{2WD}} + \frac{\sqrt{2WD - 3}}{\sqrt{2WD}} \tag{12}$$

Replacing the appropriate term in Equation 11 by the approximation in Equation 12, and solving for $U_{1-\alpha}$ we get:

$$U_{1-\alpha} \doteq \frac{\hat{\phi}\sqrt{2WD} - \phi_1\sqrt{2WD - 3}}{\phi_1 + \sqrt{2}} \tag{13}$$

Equation 13 is the confidence level of the negation of one or more overexposures, i.e., zero overexposure and similarly the negation of zero overexposure becomes:

$$U_{1-\alpha} = \frac{\hat{\phi}\sqrt{2WD} + \phi_0\sqrt{2WD - 3}}{\phi_0 + \sqrt{2}} \tag{14}$$

EXAMPLE

A zone in a workplace contained 40 workers, and the workers were exposed to compounds with threshold limit values (TLV) of 100 ppm over an eight-hour period. We were willing to accept a distinct probability of one excess once every ten days. The worker-days value in this example is 400, and if we arbitrarily choose the probability of one exposure to be less than 1% in 40 worker-days (any given day) from Figure 2 we obtain ϕ_0 to be 2.98. Similarly, if we choose the tolerable level of one exposure occurring in 400 worker days or 10% probability of zero exposure at 400, we get ϕ_1 to be 2.53, and thus we set the limits of decision. In this zone, seven workers were chosen for sampling and sampled for two days. The results of sampling are shown in Table I. It should be noted that at least one result is quite close to the TLV, and calculated $\hat{\phi}$ is greater than ϕ_0 assigned for the decision. If we calculate the confidence limits for $\hat{\phi}$, we observe that one of these limits fall within indecision range. In the strategy proposed here this does not play an important role. Because we calculate the confidence level of the decision by Equation 13:

$$U_{1-\alpha} = \frac{3.13851 \sqrt{28} - 2.53 \sqrt{25}}{2.53 + \sqrt{2}} = 1.003$$

which suggests a P value of 0.84. Under these circumstances we have two choices: (1) to accept the decision perhaps not as confidently as we wish, or (2) to decide to gather additional samples to improve the estimate. Here we note that we have collected nearly the minimum number of samples and we still have a significant number of sampling days (at least three more days) before we reach the point of diminishing returns. Had the results obtained here been from a sample containing about 40 WD, then we would not expect to improve the estimate significantly by further sampling. Under this condition we would have to make a decision based on the results obtained.

Table I. Results of Surveying Seven Workers for Two Days

Raw Data		
	C^a (ppm)	
Worker	Day (1)	Day (2)
1	29.87	29.87
2	34.62	17.60
3	44.09	64.15
4	23.95	23.21
5	48.36	53.89
6	33.13	20.92
7	93.73	38.91
Analysis of Variance Table		
	SSq	df	E(S)
Worker	1.95227	6	0.325378
Days	0.80603	7	0.115147

[a]Mean lnC = 3.58657; $\hat{\phi}$ = (ln 100 − ln C)/0.32455 = 3.13851.

REFERENCES

Corn, M., and N. A. Esmen (1979) "Workplace Exposure Zones for Classification of Employee Exposures to Physical and Chemical Agents," *Am. Ind. Hygiene Assoc. J.* 40:47.

Drinker, P., and T. F. Hatch (1936) *Industrial Dust* (New York: McGraw-Hill Book Company), p. 88.

Esmen, N., and Y. Y. Hammad (1977) "Log-Normality of Environmental Sampling Data," *J. Environ. Sci. Health*, A12(1&2):29.

Roach, S. A. (1966) "A More Rational Basis for Air Sampling Programs," *Am. Ind. Hygiene Assoc. J.* 27:1.

Roach, S. A. (1977) "A Most Rational Basis for Air Sampling Programs," *Ann. Occup. Hygiene* 20:65.

CHAPTER 23

STATISTICAL EVALUATION OF SAMPLING
STRATEGIES FOR COAL MINE DUST

J. D. Bowman, K. A. Busch and S. A. Shulman

National Institute for Occupational Safety and Health
Centers for Disease Control
Cincinnati, Ohio

ABSTRACT

A statistical model for the variation in the concentration of dust at
a sampling site in a mine has been developed and used to assess sampling
strategies for determining a coal mine's compliance with occupational
health standards. An example of this model was used to evaluate alternative
sampling strategies for the Mine Safety and Health Administration (MSHA)
to sample respirable dust in surface coal mines.

INTRODUCTION

In 1980 MSHA proposed regulations for sampling respirable dust in
surface coal mines and the surface works of underground mines [MSHA
1980a]. These proposed regulations included a sampling strategy prescribing
when and where coal mine operators should take dust samples. Criteria
were also included for deciding whether the standard for respirable coal
mine dust had been exceeded at a sampling site during the period of
sampling.

The National Institute for Occupational Safety and Health (NIOSH)
evaluated MSHA's proposed sampling strategy using a log-normal model
fitted to variations in the dust concentrations measured by MSHA in

coal mines [Busch 1980]. Quantitative statistical measures for the effectiveness of MSHA's proposed strategy were established on the basis of the modeling results. An improved sampling strategy was developed and evaluated by the same statistical measures before being recommended to MSHA in NIOSH's testimony on the proposed regulations [NIOSH 1980]. MSHA has since implemented final regulations with a third sampling strategy different from either proposal [MSHA 1980b].

In the course of this analysis, we became aware that the log-normal model was not adequate to study the reliability of consecutive shift samples for determining compliance over a two-month period. Therefore, we began to develop more sophisticated models based on a time series analysis of MSHA's dust concentration measurements in underground coal mines.

In this chapter we present an autocorrelation model fitted to some of MSHA's data for coal mine dust. We also present the results of our 1980 evaluation of the strategies for surface coal mines, which was done under a simpler log-normal model that did not include autocorrelation. This application illustrates a general method for evaluating sampling strategies, which can also be used with the more sophisticated models under development.

THE STATISTICAL MODEL

Leidel and Busch [1975] and Leidel et al. [1975] concluded that a log-normal model is adequate for describing the random variability in several types of industrial hygiene data. The variability may be viewed as a combination of environmental fluctuations distributed log-normally and as sampling and analytic error distributed normally. The environmental variability usually exceeds the random errors due to sampling and analytical errors, so that the total variability can be approximated well by a log-normal distribution. Random variability is at times superimposed about a time trend.

These concepts may be formalized mathematically as follows. Let $X(t)$ denote the time-weighted average (TWA) concentration of respirable dust at a specific location in a coal mine on day t. Then, the log-transformed concentration $Y(t)$ is given by:

$$Y(t) = ln \ X(t)$$

$$Y(t) = T(t) + E(t) + S(t)$$

where $T(t)$ = systematic time trend in the series of concentrations
    $E(t)$ = environmental variations about the trend
    $S(t)$ = random sampling and analytic errors

E(t) denotes a random nontrend component, which has a zero mean value and is normally distributed (in the logarithms). Autocorrelation may exist in the component $E(t)$ and be reflected as autocorrelation in values of $Y(t)$. The sum $T(t) + E(t)$ represents the true concentration on day t. When the model is estimated from actual measurements as in the example presented below, the random components in $E(t)$ cannot be separated from the sampling and analytical errors $S(t)$. When we are unable to separate $E(t)$ and $S(t)$, we define $Z(t) = E(t) + S(t)$ and write $Y(t) = T(t) + Z(t)$.

The trend $T(t)$ could be any systematic time trend in the series of concentrations, associated with process factors, seasonal environmental changes and other factors that cause long-term changes in the true concentration. $T(t)$ may be a constant, a linear function of t, a step function, etc., but cannot have zero mean value. When $T(t)$ is constant, the model becomes one for random variability with no trend.

In a more complex example, we would expect a downward trend in the daily TWA of dust in a mine that is out of compliance, but that is making active efforts to reduce the dust concentrations. Looking at the data in the second panel of Figure 1, we can see such a trend. We might suppose that the trend is linear and so represent $T(t) = a + bt$. Although it is difficult to attach a physical meaning to a and b, note that a linear downward trend (i.e., negative b) in the logarithms corresponds to an exponential downward trend in the concentrations themselves. If active efforts are being made to lower the dust concentrations in the work area, we would expect a decrease in the dust concentrations, which would eventually achieve a state of statistical control.

The short-term deviations of the data points from the linear trend may exhibit an autocorrelation, which measures how closely related a deviation from the trend on a given day is to a deviation for the day before. If it is above the trend line on day $t - 1$, perhaps it will tend to also be above on day t. A specific mathematical model embodying such an effect is:

$$E(t) = cE(t - 1) + \epsilon(t)$$

where $-1 < c < 1$ is the autocorrelation coefficient and $\epsilon(t)$ represents normally distributed random variations.

COAL MINE DUST DATA

MSHA has provided a computer tape of the dust samples taken by underground coal mine operators for monitoring compliance with the

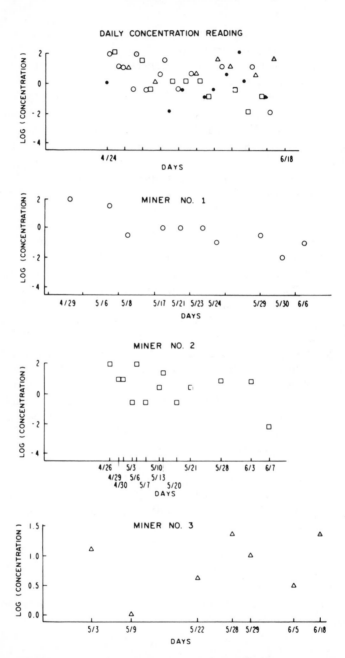

Figure 1.   Series of respirable dust samples from 8-hr shifts on a single section of an underground coal mine. Top graph shows all samples for the continuous miner operator (high-risk occupation). The other graphs show the samples for each miner who worked at that occupation. Source: Mine Safety and Health Administration records.

dust standards. These records include samples from both continuous and longwall sections. For each sample, the coal mine, the section, the type of mining, the miner being sampled and the miner's occupation are labeled with a code number.

Initial and final filter weight and sampling time information were also provided. These were used to compute dust concentrations as follows:

$$\text{concentration} = \frac{(\text{final filter weight} - \text{initial weight}) \times 1.38}{\text{sampling time} \times \text{flowrate}}$$

where the flowrate is 2 liter/min, and 1.38 is the respirable dust conversion factor used by MSHA [Tomb et al. 1973].

For time series analysis, we need long strings of consecutive samples. Under MSHA's regulations in this period before 1980, most sampling was done in strings of 5 or 10 consecutive shifts. Longer series of consecutive samples occurred only when a mine section was found to be in violation of the health standards. Under those conditions, sampling of the "high-risk occupations" (e.g., continuous miner operators) was required during each production shift until the sections' dust concentrations complied with the standard. During these long strings, improvements in the dust controls were presumably being made so that downward trends in the concentration can be expected.

Before the autocorrelation analysis, a sample string of 39 readings on continuous miner operators was first examined. This made us aware of some of the problems with the data and of the conclusions that can be drawn from the available data. The first panel of Figure 1 shows a plot of these 39 readings on continuous miner operators from the same mine section. The readings were made between April 24 and June 18, 1974. Note that this sequence of samples contains many gaps over time. On weekends and other occasions unexplained on the data tape, no mining and therefore no sampling occurred. In analyzing this sample series, we assume that these gaps can be ignored.

Of the 39 readings, 29 samples were taken on just three miners. The data for these three miners are shown separately in the three lower plots of Figure 1. Looking at these plots, several observations are in order:

1. Even though the data concern workers in the same mine section, there appears to be considerable variation from miner to miner. Since the individual mining machine operator apparently has a significant effect on dust concentration, the series of dust samples should be analyzed separately for each worker.
2. The only series that appears to have a definite trend is the one for the first miner, and the trend there is clearly downward. In this case more than five consecutive readings were taken when a mine section was

out of compliance and corrective measures seem to have been undertaken. There is no reason that the behavior of such a series should be identical to a series of point readings when the mine is still in compliance.

3. Despite the shortcomings in the MSHA data, it is the best data set currently available for studying the time series behavior of coal mine dust concentrations. Therefore, we have used the MSHA data to develop a method for estimating the parameters in the autocorrelation model for the dust concentrations.

Past studies of contaminants in a mine environment [Smaby 1978; Trimbach 1980] have found that there is an autocorrelation between the concentrations from one day to the next. Had we better information on the nature and prevalence of such relationships, we could determine sampling strategies that would tell us more easily when dust concentration in a mine was likely to go out of control. Therefore, we examined a second series of MSHA dust samples in more detail to indicate better the value of this approach.

TIME SERIES ANALYSIS

The series of dust samples plotted in Figure 2 are from a continuous mine section during February and March 1974. Several remarks are in order concerning construction of the series. In the original data, two samples are reported for each date. The second reading was usually taken during the second shift and is treated that way in the time series analysis. The value reported for February 20 appears to be an outlying value.

Outliers are found frequently in coal mine dust data, due to poor quality control or human interference with sampling [GAO 1975]. Since an outlier can significantly reduce the estimate of the autocorrelation coefficient from a short sequence, a system for removing them is crucial to this analysis. Because the outlier in Figure 2 does not seem to follow the trend of the remainder of the series, we arbitrarily deleted that value.

The concentrations were transformed to logarithms before performing the time series analysis and the logs (with the third point deleted) are shown in the second panel of Figure 2. As was discussed above, we are interested in estimating the trend $T(t)$ and the time series autocorrelation for the residuals $Z(t)$ in the model:

$$Y(t) = T(t) + Z(t)$$

Noticing the downward trend in Figure 2, the data appear to change over time. Before we can fit an autoregressive model, we remove the time-dependent portion of the trend by differencing the data:

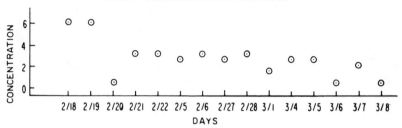

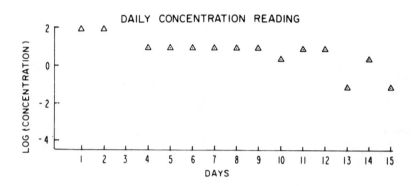

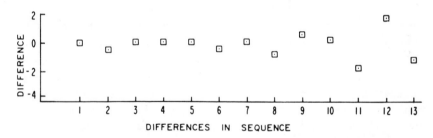

Figure 2.   Time series analysis on a sequence of respirable dust samples. Top: dust concentration from a series of consecutive-day samples on a single section of an underground coal mine. Middle: log transform of the series with one outlier removed. Bottom: differencing the series to remove the downward trend.

$$\Delta Y(t) = Y(t) - Y(t - 1) = Z(t) + \text{constant}$$

Next the differences are fit to a first-order autocorrelation model:

$$\Delta Y(t) = \overline{\Delta Y} + c(\Delta Y(t - 1) - \overline{\Delta Y}) + \epsilon(t)$$

where   c = autoregressive coefficient
           $\epsilon(t)$ = normally distributed random deviations

The differenced data $\Delta Y(t)$ in Figure 2 display no particular trend, so we next fit the above autoregressive model to the differenced data. Based on 12 pairs of $\Delta Y(t)$, the estimate of c is -0.78. The constant $\overline{\Delta Y}$ is -0.22, which is not different from zero at the 95% confidence level. Therefore, the model is:

$$\Delta Y(t) \cong -0.8 \, \Delta Y(t-1)$$

or

$$Y(t) \cong 0.2 \, Y(t-1) + 0.8 \, Y(t-2)$$

The model accounts for about 50% of the variation in the data, and the estimate of c is significantly different from 0 at the 0.05 significance level. However, with so few data points, it is uncertain whether or not this model is an adequate description of the data. The results merely serve to illustrate how an autoregressive model might be fitted to data such as MSHA has provided.

STRATEGY EVALUATION

An evaluation of sampling strategies for surface coal mines was conducted in 1980 using a simpler log-normal model in which trends and autocorrelation are neglected. MSHA had proposed regulations for dust sampling at "designated work positions" in surface coal mines and in the surface works of underground mines [MSHA 1980a]. For any work position (drag-line operators, preparation plant occupations, etc.), the proposed strategy dictates how many samples to take, and when to issue citations for violation of the health standards for respirable coal mine dust.

As outlined in Table I, the strategy proposed by MSHA consisted of three stages: an initial designation of a site for sampling, a bimonthly sampling cycle and the decision to end sampling if the dust levels stay in compliance at the work position. The bimonthly sampling cycle itself consists of a single screening sample, and if necessary, five more compliance samples. Under the MSHA proposal, the work position would have been declared in violation only if the average of those five respirable dust samples exceeds the standard (2.0 mg/m^3 for dust with less than 5% silica). Then, MSHA reviews the dust control measures and takes other steps to assure future compliance.

Table I. Sampling Strategies for Respirable Dust at Work Positions
in Surface Coal Mines

	MSHA Proposal	NIOSH Proposal	MSHA Final Rule
Initial Designation	Position designated if at least 1 sample >2.0 mg/m^3	Position designated unless 7 samples have been taken and all <2.0 mg/m^3	Position designated if at least 1 sample >1.0 mg/m^3
Bimonthly Sampling Cycle	1 screening sample >2.0 mg/m^3 implies 5 more compliance samples	1 screening sample >1.0 mg/m^3 implies 5 more compliance samples	Same as MSHA original proposal
End of Designation	3 cycles (6 months) in compliance	Same as MSHA original proposal	6 cycles (12 months) in compliance

To evaluate the effectiveness of this complex proposal for a sampling strategy, we first assumed that the single-shift time-weighted average dust concentrations are independently distributed (no autocorrelation) and can be represented by a log-normal distribution. The parameters of the log-normal distribution are the geometric mean (GM), assumed to be characteristic of each work position, and the geometric standard deviation (GSD), assumed to be constant for all work positions.

For this model, the GSD was derived from a 1974 statistical analysis of dust concentrations in underground (not surface) coal mines done by MSHA. A selection of 40 dust samples taken at the high-risk occupations in each of 22 mining sections gave a pooled coefficient of variation (CV) = 0.916. The CV was derived from the arithmetic standard deviation and arithmetic mean for each mining section. Assuming a log-normal distribution, its GM and GSD can be calculated from the CV and the long-term arithmetic mean μ by:

$$GSD = \exp \sqrt{ln(1 + CV^2)}$$

$$= 2.183$$

$$GM = \mu \exp(-ln^2 GSD/2)$$

$$= 0.737\,\mu$$

Knowing an estimate for CV, the frequency distribution of the dust concentrations at a work position can be expressed as a log-normal distribution dependent only on μ.

For a chronic disease such as coal worker's pneumoconiosis (CWP),

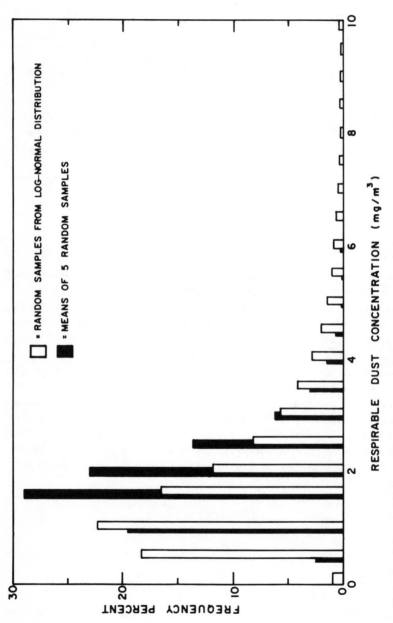

Figure 3.   Example of the concentration distributions generated by Monte Carlo simulation. In this example, the long-term average concentration $\mu = 2.0$ mg/m^3 (GM = 1.47 mg/m^3) and CV = 0.916 (GSD = 2.18).

the long-term arithmetic mean μ is the best indicator of health risk, because it is proportional to the dose of dust deposited in a miner's lung. In fact, epidemiological studies [Hurley et al. 1979; Reisner and Robock 1977] have identified the dose as the cause of CWP. In this evaluation, we shall therefore assume that a good sampling strategy will have a high probability of detecting mining sections with μ greater than the health standard.

Using the log-normal distribution derived above, the probabilities can be calculated for the three possible outcomes of a bimonthly sampling cycle, shown in Table II. The probabilities of the concentration X from a single sample for a given value of μ can easily be derived from tabulations of the cumulative standard normal distribution. However, the frequency distributions for the arithmetic means $\overline{X}$ of five samples from a log-normal distribution were calculated by numerical methods based on Monte Carlo sampling.

Figure 3 shows the X and $\overline{X}$ distributions resulting from 50,000 random samples drawn from a hypothetical log-normal distribution with μ = 2.0 mg/m^3 and CV = 0.916. The distribution of the means $\overline{X}$ in this plot is clearly not normal, although the skewness is substantially reduced from the parent log-normal distribution for X. This behavior, which is intermediate between the normal and log-normal distributions, makes a Monte Carlo calculation imperative to determine the frequency distribution for the averages of 5 samples.

From the Monte Carlo distributions the probability of compliance Prob[comp] and the average number of samples (N) can be calculated for a single cycle, using the formulas in Table II. To model the sampling strategies in Table I, the results in Table II have to be extended to three or six sampling cycles until either sampling is terminated at work positions

Table II. Outcome for a Bimonthly Sampling Cycle[a]

Case	Screening Sample	Compliance Samples	Decision	Number of Samples
1	X ≤ SL	None	Compliance	1
2	X > SL	$\overline{X}$ ≤ PEL	Compliance	6
3	X > SL	$\overline{X}$ > PEL	Violations, dust controls improved	6

Prob[compliance] = Prob[case 1] + Prob[case 2]
$$= \text{Prob}[X \leq SL|\mu] + \text{Prob}[X > SL|\mu]\,\text{Prob}[\overline{X} \leq PEL|\mu]$$
Average number of samples = (N)
$$= \text{Prob}[X \leq SL|\mu] + 6\text{Prob}[X > SL|\mu]$$

[a]Notation: X = dust concentration for a single shift; $\overline{X}$ = arithmetic average for five shifts; SL = screening level; PEL = permissible exposure limit (health standard).

decided to be in compliance or a citation is issued at positions decided to be in violation. (In the latter case, the dust control plan will presumably be altered so that the long-term average μ will be shifted to a lower value.) The general formulas are:

$\text{Prob}_n[\text{comp}]$ = probability of deciding for compliance after the set of n bimonthly cycles (n = 3 or 6)

$$= (\text{Prob}[\text{comp}])^n$$

$\text{Prob}_n[\text{noncomp}] = 1 - (\text{Prob}[\text{comp}])^n$

$(N)_n$ = expected number of samples taken in n bimonthly cycles

$$= (N) \sum_{i=0}^{n-1} (\text{Prob}[\text{comp}])^i$$

The results are shown in Figure 4 in terms of the probability of deciding for compliance. In Table III, the same results are expressed as the "operator's risk" (α) the probability of a citation if $\mu > 2.0$ mg/m^3 and as the "miner's risk" (β), the probability of finding the work position in compliance if

Table III. Risk Probabilities with Sampling Strategies for Surface Coal Mines

Average Dust Concentration μ (mg/m^3)	Percent of Shifts >2 mg/m^3	Risk Probabilities			
		MSHA Proposal	NIOSH Proposal	MSHA Rule	
0.8	5	0.00	0.00	0.00	↑
1.0	10	0.01	0.02	0.02	
1.2	15	0.03	0.08	0.06	Operator's risk
1.4	20	0.07	0.19	0.14	(probability of
1.6	25	0.15	0.32	0.28	noncompliance)
1.8	30	0.26	0.49	0.45	
2.0	35	0.38	0.64	0.62	
2.0	35	0.62	0.36	0.38	
2.2	39	0.50	0.24	0.25	
2.4	44	0.40	0.15	0.16	
2.6	48	0.31	0.10	0.10	
2.8	52	0.23	0.06	0.05	Miner's risk
3.0	55	0.17	0.03	0.03	(probability of
3.5	63	0.08	0.01	0.01	compliance)
3.7	65	0.06	0.00	0.00	
4.0	69	0.04	0.00	0.00	
4.5	74	0.02	0.00	0.00	
5.0	78	0.01	0.00	0.00	
5.8	73	0.00	0.00	0.00	↓

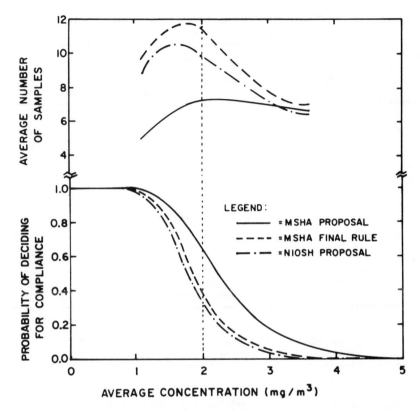

Figure 4. Probability of deciding for compliance and the average number of samples for the different sampling strategies compared to the 2.0-mg/m³ standard for respirable coal mine dust (silica <5%).

$\mu < 2.0$ mg/m³. In designing sampling strategies, NIOSH aims at keeping these risks α and β as low as possible, with 0.05 being the goal for α in legally binding decisions. The long-term average dust concentrations that correspond to low risk levels are shown in Table IV. Finally, the average number of samples taken to reach a decision is given as a function of μ in Table V. These results apply only to the final two stages of the sampling strategies. The operating characteristics of the initial designation stage in Table I are analyzed separately by Busch [1980].

Under MSHA's proposed strategy, a work position where μ equals the 2.0-mg/m³ standard would have a 62% probability of being found in compliance. The average dust concentration would have to be at least 3.9 mg/m³, 195% of the standard, for the miner's risk to be no greater

Table IV. Long-Term Average Dust Concentrations (mg/m^3)
Required to Keep the Risks Low

Selected Maximum Risk Levels	Average Dust Concentrations		
	MSHA Proposal	NIOSH Proposal	MSHA Rule
Miner's Risk (β)			
0.10	3.4	2.6	2.6
0.05	3.9	2.9	2.8
0.01	5.1	3.4	3.3
Operator's Risk (α)			
0.10	1.5	1.2	1.3
0.05	1.3	1.1	1.2
0.01	1.0	0.9	0.9

Table V. Average Number of Samples for Surface Coal Mine
Sampling Strategies

Average Dust Concentration μ (mg/m^3)	Average Number of Samples		
	MSHA Proposal	NIOSH Proposal	MSHA Rule
1.1	4.8	8.8	9.5
1.3	5.5	9.7	10.7
1.9	7.0	10.1	11.7
2.0	7.1	9.9	11.5
2.3	7.2	9.0	10.4
2.7	7.1	8.0	9.0
2.9	7.1	7.6	8.5
3.9	6.6	6.4	6.9

than 5%. The probability of escaping a citation ranges from 5% at μ = 3.9 mg/m^3 to 62% at μ = 2.0 mg/m^3.

Believing that the MSHA proposal allowed an unacceptably high level of risk for the miner's health, NIOSH [1980] recommended some minor changes in the MSHA strategy (Table I), such as reducing the screening level from 2.0 to 1.0 mg/m^3. These changes clearly reduced the miner's risk (Figure 4 and Table III). MSHA, however, took the position that the NIOSH proposal "would result in more sampling than MSHA believes is necessary" [MSHA 1980b], and adopted a third strategy for their final regulations (Table I).

According to this analysis, the miner's protection against long-term average dust concentrations exceeding the standard are essentially equal under both the NIOSH proposal and MSHA's final rule. Since MSHA had raised the issue of the number of samples required under the various proposals, we also calculated the average number of samples for MSHA's final rule. Ironically, MSHA's final strategy requires more samples on average than the NIOSH proposal. This outcome demonstrates that qualitative arguments can be unreliable in judging the relative costs and benefits of complex sampling strategies.

On the positive side, the 12-month requirement in MSHA's final rules eliminates any seasonal bias in the dust concentration, an important consideration in surface mines. From a public health viewpoint, we therefore consider MSHA's final rule to be the best strategy of the three.

CONCLUSIONS

Statistical evaluation is an objective technique for analyzing sampling strategies, both for their effectiveness in assuring compliance with health standards and for their costs in sampling resources. Quantitative evaluation of the sampling strategies for surface coal mines demonstrated the need for improvements in the regulations originally proposed by MSHA. The rules finally adopted by MSHA provide protection to the miner equal to NIOSH's alternative, while eliminating seasonal bias in the sampled dust concentrations.

The autocorrelation models described here are presently being used to improve further the effectiveness of industrial hygiene sampling. To develop reliable autocorrelation models, we should clearly deal with longer series of data points. The statistical parameters should also be derived from data taken at more mines and occupations for the model to be representative of the conditions in U.S. coal mines.

We plan to investigate the data on MSHA's dust sample records from longwall sections where longer series of consecutive samples historically have been taken. All of the longer sequences on the MSHA data tape are presumably biased by efforts to bring the mine section into compliance with the health standards. Field studies may be necessary to estimate autocorrelation in an unbiased environment.

If appreciable autocorrelation is present, the resulting statistical model will allow us to estimate more efficiently the mean dust concentration. With or without correlation, the time series methods described here are useful in the development of more sophisticated methods for the monitoring and control of dusts in mines.

ACKNOWLEDGMENTS

The authors wish to thank the members of the NIOSH staff who contributed to this research: Randall Smith, Teresa Trimbach and Alan Armstrong performed much of the computer calculations; August Lauman prepared the graphics; and Roberta Crutchfield typed the manuscript.

REFERENCES

Busch, K. A. (1980) "Statistical Evaluation of Title 30. CFR, Part 71, MSHA Coal Mine Dust Sampling Regulations," NIOSH memorandum.

GAO (1975) "Improvements Still Needed in Coal Mine Dust-Sampling Program and Penalty Assessments and Collections," Report to Congress by the Comptroller General of the United States No. RED-76-56.

Hurley, J. F., L. Copland, J. Dodgson and M. Jacobsen (1979) "Simple Pneumoconiosis and Exposure to Respirable Dust: Relationships from Twenty-five Years' Research in British Coal Mines," Institute of Occupational Medicine Report No. TM/79/13.

Leidel, N. A., and K. A. Busch (1975) "Statistical Methods for the Determination of Noncompliance with Occupational Health Standards," NIOSH Technical Publication No. 75-159.

Leidel, N. A., K. A. Busch and W. E. Crouse (1975) "Exposure Measurement Action Level and Occupational Environmental Variability," NIOSH Technical Publication No. 76-131.

MSHA (1980a) "30 CFR 71, Respirable Dust, Proposed Rule and Notice of Public Hearings," *Federal Register* 45:24009.

MSHA (1980b) "30 CFR 71, Respirable Dust, Operator Sampling Procedures, Final Rule," *Federal Register* 45:80746.

NIOSH (1980) "NIOSH Comments on MSHA's Proposed Rules for Respirable Dust at Surface Coal Mines (30 CFR 71) . . . ," Testimony before MSHA Public Hearings.

Reisner, M. T. R., and K. Robock (1977) "Results of Epidemiological, Mineralogical, and Cytotoxicological Studies on the Pathogenicity of Coal-Mine Dusts," in *Inhalable Particles IV*, W. H. Walton, Ed. (Oxford: Pergamon Press, Ltd.).

Smaby, S. A. (1978) "The Development and Application of Statistical Air Quality Data Analysis Techniques," MS Thesis, Michigan Technological University.

Tomb, T. E., H. N. Treaftis, R. L. Mundell and P. S. Parobeck (1973) "Comparison of Respirable Dust Concentrations Measured with MRE and Personal Gravimetric Sampling Equipment," Bureau of Mines Report of Investigations No. 7772.

Trimbach, T. (1980) "Coal Mine Dust Study," NIOSH memorandum.

CHAPTER 24

APPLICATION OF THE APPARENT DIAMETER METHOD FOR INVERSION OF PENETRATION: ISOMETRIC PARTICLES

P. Y. Yu and J. San

Department of Chemical and Nuclear Engineering
University of Maryland
College Park, Maryland

J. W. Gentry

Institute for Physical Science and Technology
University of Maryland
College Park, Maryland

ABSTRACT

Recent developments in the use of the "apparent diameter" method for the inversion of penetration measurements are described in this chapter. The essence of this method is to solve the integral equation:

$$\int_0^\infty F(D)Ef(D,Z)dD = \overline{Ef}\,(Z)$$

for the unknown size distribution function $F(D)$, given experimentally measured efficiencies $\overline{Ef}(Z)$ and a theoretical expression for the efficiency $Ef(D,Z)$. The "apparent diameter" method is based on the behavior of $\overline{D}(Z)$ defined by:

$$Ef[\overline{D}(Z),Z] \equiv \overline{Ef}(Z)$$

In this chapter it is established that the conjecture:

$$\int_0^1 ln\overline{D}(Z)d\overline{Ef} = \int_0^\infty lnDF(D)dD$$

is valid for all power law classifiers, and the use of the method in analyzing experimental measurements is described, and the method is critically compared to procedures based on the Nyström method.

INTRODUCTION

In this chapter, the use of the "apparent diameter" method to characterize multimodal particle size distributions is described. To illustrate the method, we consider power law classifiers—those in which the collection efficiency can be described by relations of the form:

$$Ef(D,Z) = \left[\frac{D}{D(Z)}\right]^n \quad \text{for } D \leqslant D(Z)$$

$$= 1 \qquad \text{for } D > D(Z) \tag{1}$$

where $D(Z)$ is a characteristic diameter—a function of controllable variables Z (e.g., flowrate, residence time and classifier dimensions). The exponent n is characteristic of the type of classifier. Furthermore, it is assumed that the particle size distribution can be characterized by a sum of log-normal distributions. Specifically, the particle size distribution is given by:

$$F(D) = \sum_j \frac{C_j}{\sqrt{\Pi}\,\sigma_j} \exp -\left[\frac{ln^2 D/D_j}{\sigma_j^2}\right] \frac{dlnD}{dD} \tag{2}$$

where $D_j, \sigma_j/\sqrt{2}$ = mean diameter and the standard deviation, respectively, of the constituent distributions
C_j = relative weights of the distributions

By restricting the investigation to power law classifiers and log-normal distributions, the underlying assumptions behind the method can be examined. In using the apparent diameter method, it is necessary to invert Equation 1, expressing the apparent diameter as a function of efficiency. If the efficiency were a power law, then the inverse can be expressed analytically—a necessary condition to obtain analytical expressions. The use of the method itself is not so constrained; the principal requirements are that the function is monotonic (either increasing or decreasing) and that for some diameter, the entire range of efficiencies between 0 and 1 is covered. By assuming a series of log-normal distributions, one can approximate almost all feasible (nonnegative everywhere) size distributions.

Three basic questions were examined:

1. What can one conclude about the effect of polydispersity on the solution technique? Of particular interest is the determination of distribution parameters from incomplete data.
2. Can an essential step in the procedure—the conjecture that:

$$\int_0^1 ln \; \overline{D} d\overline{Ef} = \int_0^\infty ln \; DF(D)dD \qquad (3)$$

   where $\overline{D}$ is the apparent diameter and $\overline{Ef}$ is the efficiency—be proved subject to constraints on the functional form of the efficiency and the distribution function?
3. Can one develop an efficient algorithm for cases where the distribution is multimodal?

The potential advantage of the "apparent diameter" method lies in that there is less ambiguity in the parameters—for a single log-normal distribution, the parameters are deterministic. It is very convenient to use and requires fewer data points to estimate the parameters. Although the development below is for power law classifiers, many real classifiers— diffusion batteries, elutriators, inertial impactors, cyclones and settling chambers—can be approximated by power law classifiers that show correct behavior at the asymptotic limits. For example, diffusion batteries may be simulated by $n \sim 2/3$, elutriators or gravitational sedimentation by $n \sim 2$, cyclones or centrifuges by $n \sim 2$–2.5, and inertial impactors by $n > 4$.

PREVIOUS WORK

Research in the analysis of particle size distributions has, for the most part, concerned three specific problems: (1) interpretation of diffusion battery measurements; (2) correction of inertial impactor measurement for nonideal stages; and (3) unambiguous definition of multimodal distributions. In essence, all the problems reduce to the solution of an integral equation:

$$\overline{Ef}(Z) = \int_0^\infty F(D)Ef(D,Z)dD \qquad (4)$$

where $\overline{Ef}$ = experimentally measured efficiency
   $Ef(D,Z)$ = theoretical efficiency in terms of particle diameter D and controllable variable Z
   $F(D)$ = distribution function

Generally, the expression for the theoretical efficiency is known and the experimental efficiency can be measured within 1-5%. The computational problem is to determine the distribution function F(D).

Most previous work has focused on the analysis of diffusion battery measurements. Efficiency measurements are made at a number of different flowrates or with a diffusion battery containing sections of different lengths. The theoretical efficiency is a known function of particle size; consequently, determining the size distribution reduces to the solution of Equation 4. Often, an approach suggested by Mercer and Greene [1974] is used: employing the diffusion coefficient (a monotonically decreasing function of diameter) is used, rather than particle diameter as the independent variable in Equation 4. This approach is more convenient, because theoretical efficiency is an explicit function of the diffusion coefficient approaching a power law relation (n ~ 2/3) as the efficiency approaches zero [Gormley and Kennedy 1949].

Previous investigators have attempted to solve Equation 4 by three methods:

1. The Nyström [1930] method: the integral is replaced with a quadrature expression. The resulting algorithm is a set of linear algebraic equations with the numerical values of the distribution function being the unknown variables. (Related methods are also used in which the kernel of Equation 4 is modified to produce a better conditioned matrix [Maigne et al. 1974; Perrin 1980].

2. The invariant function form (IFF) method: a specific functional form (i.e., log normal distribution) is assumed [Fuchs et al. 1962]. The distribution parameters which give the best fit are obtained by a suitable minimization algorithm [Raabe 1978; Soderholm 1979]. Two criteria must be specified: a specific functional form of the distribution, and the minimization algorithm, usually based on comparison of the efficiency measurements with simulated efficiencies using the trial distribution.

3. The "apparent diameter" method [King et al. 1980; Park et al. 1980; Pollak and Metnieks 1957]: the distribution is obtained by analyzing the behavior of the apparent diameter as a function of efficiencies or penetration. The apparent diameter $\overline{D}$ is defined by the relation: $Ef(Z) = Ef(\overline{D}, Z)$; that is, the apparent diameter $\overline{D}$ is the hypothetical diameter of a monodisperse distribution that has the same efficiency at conditions Z as the distribution.

Each of the approaches has limitations. Only the Nyström methods do not assume a priori a specific functional form. Unfortunately, one rarely has measurements of sufficient precision that one can obtain an unambiguous distribution. What frequently happens is that the resulting matrix (based on the numerical quadrature of the right side of Equation 4) is so badly conditioned that very small measurement errors result in nonfeasible values. As a result, the IFF method is the most widely used procedure.

Aside from the restriction imposed by assuming a specific functional form, the IFF approach suffers from two additional problems: (1) the minimization equations are nonlinear and (2) there is no assurance that a global rather than a local minimum is found. In short, the results are ambiguous and the algorithm often is tedious to apply. We believe that a better method is based on the use of the apparent diameter. The simplest example of this method [Megaw and Wittin 1963; Nolan and Scott 1963] was to assume that the apparent diffusion coefficient at a specific efficiency (usually 60-65%) equals the log mean diffusion coefficient—one of the two parameters in the log-normal distribution. This simple assumption worked very well, but it provided no information regarding the second parameter, standard deviation. A second approach was to examine asymptotic behavior as the efficiency approaches one [Pollak and Metnieks 1957]. In this case, the distribution was assumed to consist of the sum of monodisperse distributions—an idea similar to that of the Nyström approach and not restricted to a functional form assumed a priori. The method did not prove useful principally because of its sensitivity to experimental error. Park et al. [1980] and King et al. [1980], examined the asymptotic behavior of the apparent diffusion coefficient as the efficiency approaches zero. They found that:

$$\lim_{\overline{Ef}\to 0} \overline{D}*^{2/3} = \Sigma C_i D_i^{*2/3} \exp\left[\frac{\sigma_i^2}{9}\right] \tag{6}$$

where D^* = diffusion coefficient
D_i^* = log mean diffusion coefficient

$\sigma_i/\sqrt{2}$ = standard deviation

If one could find D_i^* or σ_i independently, the parameter could be obtained unambiguously.

Based on an examination of a number of simulations, Park et al. [1980] conjectured that:

$$\int_0^1 ln\ \overline{D}*d\overline{Ef} = \int_0^\infty ln\ D*F(D*)dD* \tag{7}$$

which is analogous to Equation 3. This approach proved to be successful.

DEVELOPMENT

If the particle size distribution were log-normal, and if the efficiency were a power law, then one has:

$$\overline{Ef}(Z) = \int_0^\infty Ef(D,Z)F(D)dD$$

$$= \int_0^{D(Z)} \left[\frac{D}{D(Z)}\right]^n F(D)dD + \int_{D(Z)}^\infty F(D)dD \qquad (8)$$

$$= \frac{1}{2}\exp(\overline{q}_1^2 - 2\overline{q}_1\tau_1)[1 + erf(\tau_1 - \overline{q}_1)] + \frac{1}{2}erfc(\tau_1) \qquad (9)$$

where $\quad \tau_1 = (1/\sigma_1) \ln [D(Z)/D_1]$ \hfill (10)

and $\quad \overline{q}_1 = n\sigma_1/2$ \hfill (11)

erf(y) and erfc(y) are the error function and complementary error function respectively. They are defined by $erf(y) = 2/\sqrt{n} \int_0^y e^{-t^2}dt$ and $erfc(y) = 1 - erf(y)$.

The overall efficiency can therefore be characterized by two parameters: (1) the product of the classifier exponent and the standard deviation and (2) τ, which is a normalized characteristic diameter. The parameter τ tends toward infinity when the characteristic diameter $D(Z)$ is large.

The apparent diameter can be expressed in terms of τ_1, n and $\overline{q}_1$ by:

$$\overline{D}^n = D(Z)^n\overline{Ef}$$

$$= \left[\frac{D(Z)}{D_1}\right]^n D_1^n\overline{Ef}$$

$$= \exp(2\overline{q}_1\tau_1)D_1^n\overline{Ef} \qquad (12)$$

$$= D_1^n \frac{1}{2}\exp(\overline{q}_1^2) [1 + erf(\tau_1 - \overline{q}_1)$$

$$+ D_1^n \frac{1}{2}\exp(2\overline{q}_1\tau_1)erfc(\tau_1) \qquad (13)$$

At asymptotic conditions corresponding to the efficiency approaching zero, or $\tau_1 \to \infty$, one has:

$$\lim_{\overline{Ef}\to 0} \overline{D}^n = D_1^n \exp \overline{q}_1^2 \qquad (14)$$

For the special case where the diameter D is replaced by the diffusion coefficient D* and n takes the value 2/3 (corresponding to a diffusion battery), one has

$$\lim_{\overline{Ef} \to 0} \overline{D}^{*2/3} = D_i^{*2/3} \exp \overline{q}^2 = D_i^{*2/3} \exp \frac{\sigma^2}{9} \tag{15}$$

as obtained earlier [Park et al. 1980].

The importance of this result lies in three points:

1. First, at the asymptotic condition of zero collection efficiency (penetration equal to one), the value of the apparent diameter (diffusion coefficient) can be expressed as a simple function of three parameters relating the two parameters of the log-normal distribution and the classifier exponent.
2. From Equations 9 and 13, it is possible to obtain an expression for apparent diameter as a function of efficiency. Such a plot poses the use of point-to-point comparison of calculated and measured apparent diameters rather than just the behavior at asymptotic conditions.
3. Since in the parameter $\overline{q}$, the standard deviation is multiplied by the classifier exponent, a diffusion battery with a very broad distribution behaves qualitatively as an elutriator with a narrow distribution.

To examine the effect of $\overline{q}$ on the apparent diameter–efficiency plots, the dimensionless ratio ζ:

$$\zeta = \frac{n}{\overline{q}_1^2} \ln \frac{\overline{D}}{D_1} \tag{16}$$

is plotted as a function of efficiency with $\overline{q}$ a parameter in Figure 1. As the value of $\overline{q}$ increases, the characteristic efficiency—the value where $\zeta = 0$ (the apparent diameter is equal to the mean diameter)—shifts toward 50%. This shift is shown more clearly in Figure 2, where the characteristic efficiency plotted as a function of $\overline{q}$ decreases from 1 to 0.5.

The asymptotic method of Park is based on the limit of n $\ln \overline{D}/D_1$ approaching $\overline{q}_1^2$ as the efficiency approaches zero. This is equilvanet to the statement that ζ approaches one as the efficiency approaches zero. However, nothing has been stated regarding how ζ approaches one. From Figure 2, it is evident that the slope becomes large for $q > 2$. Such behavior would make extrapolation from efficiencies of 20% and higher risky, in that too low a value of the intercept would be predicted. This error would suggest a narrower distribution than actually occurs. For small values of $\overline{q}_1$, the reverse is true, in that the slope decreases as $\overline{q}_1$ becomes small, and the method predicts a distribution more polydisperse than actually occurs. For $\overline{q}_1 = 1$, which represents conditions typical of an elutriator, one could extrapolate to zero efficiency with confidence.

Two points should be stressed. First, accurate measurements of the apparent diameter for efficiencies near zero obviate the necessity for

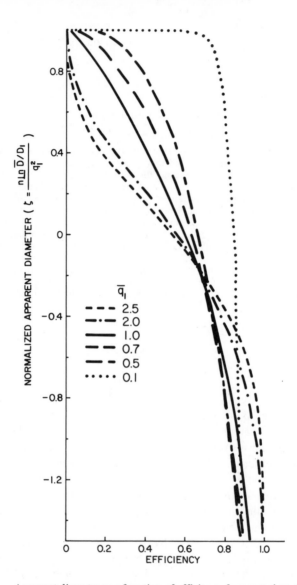

Figure 1. Apparent diameter as a function of efficiency for power law classifier.

extrapolation. Secondly, an iterative procedure can be used with extrapolations based on the values of $\bar{q}_1$ from the previous iteration to converge to the best values.

It is significant that the apparent diameter decreases precipitously for

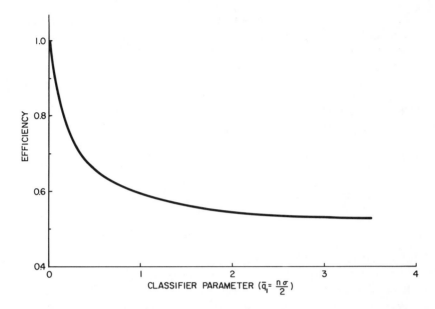

Figure 2. Characteristic efficiency where $\zeta = 0$ ($\overline{D} = D_1$) as a function of parameter $\overline{q}$.

efficiencies near one. This is, of course, an artifact of assuming a log-normal distribution; or, in fact, any distribution that has no minimum size. Nevertheless, this sharp decrease and large slope at efficiencies near one is characteristic of polydisperse distributions. Because of this behavior, methods (such as that of Pollak) dependent on accurately determining an intercept and slope as the efficiency approaches one generally fail.

As mentioned previously, the idea of assuming that ζ is equal to zero at efficiencies of 60–65% proved suprisingly good. The reason for this was that for diffusion batteries, $\overline{q}_1 \sim 0.4$, which from Figure 2 corresponds to characteristic efficiency of 0.67 near the value used in the Nolan or Megaw methods. Furthermore, although the characteristic efficiency varies with $\overline{q}_1$, it does not vary greatly. In fact, the chief source of error in the measurement lies not in uncertainty in the value of the characteristic efficiency, but in uncertainty of the value of ζ at that efficiency; for the slope of ζ with efficiency is large near the characteristic efficiency. Although this approach was developed for diffusion batteries, it should be more accurate for elutriators, in that both the slope of characteristic efficiency with $\overline{q}_1$ and apparent diameter with efficiency are flatter at higher values of $\overline{q}_1$.

Although the method above was described only for a single log-normal, the procedure can be applied to a distribution consisting of the sum of J log-normal distributions with weights C_j, standard deviations σ_j and mean diameters D_j. For this case, one has:

$$\overline{Ef}(Z) = \sum_{j=1}^{J} \frac{C_j}{2} \exp(\overline{q}_j^2 - 2\overline{q}_j\tau_j)[1 + erf(\tau_j - \overline{q}_j)] + \sum_{j=1}^{J} \frac{C_j}{2} erfc(\tau_j) \qquad (17)$$

and

$$\overline{D}^n = \sum_{j=1}^{J=0} \frac{C_j}{2} D_j^n \exp \overline{q}_j^2 [1 + erf(\tau_j - \overline{q}_j)] + \sum_{j=1}^{J} \frac{C_j}{2} \exp(2\overline{q}_j\tau_j) erfc(\tau_j) \qquad (18)$$

PARK-GENTRY CONJECTURE

By examining the asymptotic behavior, one equation relating two parameters of a log-normal distribution can be found. (It is assumed that the exponent for the classifier is known.) The problem is to find a second equation. Examining a large number of simulations of diffusion battery measurements led to the conjecture that:

$$\int_0^1 ln\ \overline{D}d\overline{Ef} = \int_0^\infty ln\ DF(D)dD$$

$$= \sum_{j=1}^{2} C_j\ ln\ D_j \qquad (19)$$

Every simulation with unimodal or bimodal distributions indicated agreement of the right and left sides of Equation 19. It is important to note that the left side consists entirely of experimentally measured terms and is completely independent of the distribution. It depends on the type of classifier. The right side is independent of experiment and classifier depending only on the distribution. We saw no reason a priori why the relation above should be valid. We also found that if a function other than lnD [i.e., $(ln$D$)^2$] were used, the relation did not hold. The numerical simulations covered a wide range of $\overline{q}_1$.

In this section, we examine the conjecture that:

$$\int_0^1 (ln\overline{D})^a d\overline{Ef} = \int_0^\infty (ln$D$)^a F(D)dD = (ln$D$_1)^a \qquad (20)$$

subject to the conditions that: the distribution consists of a log-normal distribution; and the experimental efficiency is given in terms of the apparent diameter by the power law:

$$\overline{Ef} = \left[\frac{\overline{D}}{D(Z)} \right]^n \tag{21}$$

which can be rearranged as:

$$ln\overline{D} = lnD(Z) + \frac{1}{n} ln\overline{Ef} \tag{22}$$

From the definition of τ_1, one has:

$$lnD(Z) = lnD_1 + \frac{1}{n}(2q_1\tau_1) \tag{23}$$

Substitution of these relations gives:

$$\int_0^1 [ln\overline{D}]^a \, d\overline{Ef} = \int_0^1 \left[\frac{1}{n} ln\overline{Ef} + lnD_1 + \frac{1}{n}(2q_1\tau_1) \right]^a d\overline{Ef} \tag{24}$$

When a is equal to one, this equation reduces to:

$$\int_0^1 ln\overline{D}d\overline{Ef} - lnD_1 = \frac{1}{n} \int_0^1 ln\overline{Ef}d\overline{Ef} + \frac{1}{n} \int_{-\infty}^{\infty} 2q_1\tau_1 \left(\frac{\partial Ef}{\partial \tau} \right) d\tau$$

$$= \frac{1}{n}[-1 + I(1;q)] \tag{25}$$

where:

$$I(m;\overline{q}) = \int_{-\infty}^{\infty} (2\overline{q}\tau)^m \exp(\overline{q}^2 - 2\overline{q}\tau)[1 + erf(\tau - \overline{q})] \, d(\overline{q}\tau) \tag{26}$$

It can be shown that $I(m;\overline{q})$ is constant (in fact, equal to one) only if m is equal to zero or one. Consequently, the left side of Equation 26 vanishes. For other values of m, the integral depends on $\overline{q}$.

The importance of this result is that when a = 1, one has:

$$\int_0^1 ln\overline{D}d\overline{Ef} = lnD_1 \tag{27}$$

and that for values of a other than zero or one, the integral:

$$\int_0^1 [ln\overline{D}]^a \, d\overline{Ef}$$

is a function of $\overline{q}$ and cannot be independent of the distribution. Furthermore, one can show that for a distribution consisting of a series of log-normal distributions:

$$\int_0^1 ln \, \overline{D} d\overline{Ef} = \Sigma C_j \, ln D_j \tag{28}$$

Because an arbitrary sum of log-normals is such a versatile function, it seems unlikely that the conjecture would not hold for any feasible distribution.

Extending the method to efficiencies with expressions other than power law ones presents difficulties that we were unable to resolve. Essentially, the problem is that one cannot express the inverse (i.e., express the apparent diameter as a function of efficiency) analytically. The best we have been able to do is to show that for expressions for diffusional deposition or for sedimentation in cylindrical tubes, the method regenerates the parameters for unimodal and bimodal log-normal distributions.

MULTIMODAL DISTRIBUTIONS

As mentioned above, the algorithm provides two basic equations for multimodal distributions:

$$\int_0^1 ln\overline{D}d\overline{Ef} = \Sigma C_j \, ln D_j \tag{29}$$

and

$$\lim_{\overline{Ef}\to 0} \overline{D}^n = \Sigma C_j D_j^n \exp \overline{q}_j^2 \tag{30}$$

These two equations, along with the restriction that $\Sigma C_j = 1$, provide three constraints on the distribution. Were the distribution to consist of the sum of J log-normal distributions, there would be three J variables (C_j, D_j, σ_j or $\overline{q}_j$) and three constraints, leaving $3J - 3$ degrees of freedom. Clearly, the advantage of providing unambiguous estimates of the distribution parameters can apply only to unimodal distributions.

A considerable number of simulations were carried out for bimodal distributions. The question that we examined in detail was whether the curve of apparent diameters as a function of efficiency is sufficiently sensitive that it could distinguish between two distributions chosen so that the average value of the logarithm of the apparent diameter and the asymptotic value of the apparent diameter as the efficiency approaches zero are the same. What we examined first in this chapter was how the parameters describing a log-normal distribution function could be obtained from experimental measurements. The second question is whether these parameters are adequate to describe the distribution or should the distribution consist of more than one log-normal distribution. The simulations used as a test of this approach consisted of bimodal distributions chosen such that the standard deviation of both constituents were the same with the fraction contributed by one constituent specified ($C_1 = 0.5$ or 0.1 in the simulations) and the ratio of the values $D_1^{(2)}/D_1^{(1)}$ (0.5, 0.2 and 0.1 simulations) were specified. $D_1^{(2)}$ is the smaller of the two mean apparent diameters constituting the bimodal distribution, whereas $D_1^{(1)}$ is the mean apparent diameter of the unimodal distribution.

Typical values are shown in Figures 3 and 4. In the first case, an elutriator is used as the classifier. The distribution is computed for $C_1 = C_2 = 0.5$. On the left side of Figure 3, it is evident that the three distributions are different. When the apparent diameters are plotted (on the right side of the figure) it is evident that the curves are significantly different. It is important to remember that these distributions are chosen so that the intercept and mean value of log D are the same for all three distributions. The significance of this result is that the method is sufficiently sensitive to distinguish between similar but not identical distributions.

In the second case (Figure 4), the question is examined whether one can distinguish between a bimodal distribution when the bulk of the particles are in one peak. The idea here is whether the method could be used to determine the parameters of the accumulation and/or coarse mode as well as a nucleation mode. The left side presents four log-normal distributions, with 90% of the particles in the peak corresponding to the smallest diameters (largest diffusion coefficients). For these simulations, the diffusion coefficient replaces the diameter as the independent variable. Again, it is clear that an observable difference in the frequency (left side) is manifested by a similar difference in the apparent diffusion coefficient-efficiency plots. It is clear that, unless this difference is sufficient to indicate a difference in the frequency diagrams, it will not be detectable in the apparent diameter. This may be an important limitation of the method because bimodal and trimodal distributions are normally detected through higher moments—area or volume distributions— rather than through more detail in the number frequency.

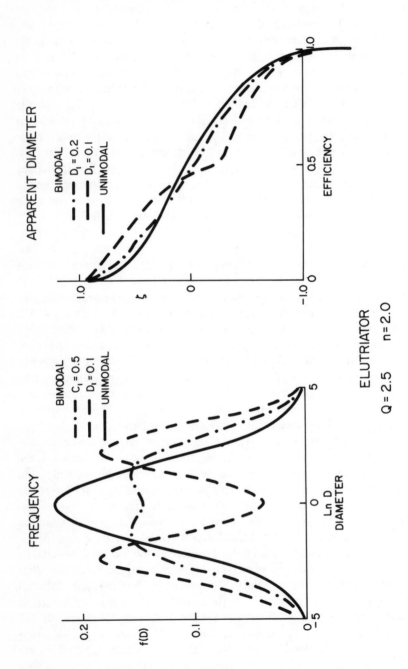

Figure 3.   Simulations of elutriator measurements for bimodal distributions.

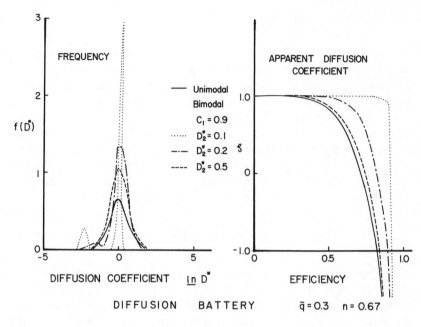

Figure 4. Simulations of measurements for bimodal distributions—diffusion battery.

EXAMPLES

The requisites for applying the method are:

1. Experimental measurements must cover the range of efficiencies from 0 to 1.
2. Theoretical efficiency must be monotonic with particle size.
3. Particle size distribution can be described by a series of log-normal distributions.

It should be emphasized that the analysis does not depend on the distribution being described by a single log-normal, nor does it depend on the theoretical efficiency having a power law dependence on the particle diameter.

However, we recommend that the data be tested first with the hypothesis that the size distribution be unimodal. To illustrate the application of the method, we consider the efficiency measurements in a settling chamber consisting of rectangular channels. The theoretical efficiency was found to be:

$$Ef = \frac{0.1 \, D^2}{Q} \qquad (31)$$

where   D = particle diameter (μ)
         Q = flowrate (liter/min)

If the channels were cylindrical, the efficiency would no longer have a simple power law form, although it would still satisfy the requirements for the algorithm and would be proportional to the square of the particle diameter in the limit as the efficiency approaches zero.

The simulated measurements are given in Table I. It is interesting to note that to cover the range of efficiencies from 5 to 95%, it is necessary to vary the flowrate over three orders of magnitude. Were n smaller, as is the case for diffusion battery measurements, the range of flowrates would be even larger. The first step in the calculation is to determine the apparent diameter. For example, Equation 31 can be rewritten:

$$\bar{D} = \sqrt{10\ \overline{\overline{Ef}}\ Q} \tag{32}$$

These values are given in column 3. The reason that the theoretical efficiency must be monotonic is to ensure that the apparent diameter is uniquely determined. The additional requirement that the efficiency be power law only allows one to write the inverse expression for $\bar{D}$ in closed form.

The next step is to calculate the logarithm of the apparent diameter and to plot these values as a function of efficiency. These data are

Table I. Simulated Efficiency Measurements for an Elutriator

Flowrate Q	Efficiency Ef	Apparent Diameter $\bar{D}$	Logarithm $ln\,\bar{D}$
0.1	1.00	1.0	0
0.2	1.00	1.5	0.38
0.5	1.00	2.2	0.77
1.0	1.00	3.2	1.15
2.2	0.98	4.6	1.53
4.6	0.94	6.6	1.89
10.0	0.86	9.3	2.23
21.5	0.66	11.9	2.48
46.4	0.55	16.0	2.77
100.0	0.37	19.2	2.95
215.0	0.21	21.0	3.05
464.0	0.12	23.3	3.15
1000.0	0.06	24.2	3.19
2150.0	0.03	24.6	3.20

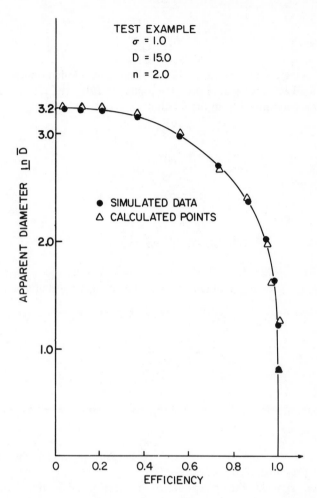

Figure 5. Application of apparent diameter method to simulated elutriator measurements.

indicated by the solid circles in Figure 5. The solid line in this figure represents the loci of a curve drawn through the data. Thus far, no assumptions have been made regarding the form of the size distribution function.

In the next step, the data are analyzed as if the size distribution were log-normal. The Park-Gentry conjecture is used to estimate the value of lnD_1. Specifically,

$$lnD_1 = \int_0^1 ln\overline{D}d\overline{Ef} \tag{33}$$

with the integral on the right side of Equation 33 determined by five-point Gauss-Legendre quadrature; the value of lnD_1 is 2.71. Second, the value of σ is determined from the relation:

$$\lim_{\overline{Ef}\to 0} [ln\overline{D}_1] - lnD_1 = \frac{\overline{q}^2}{n} = \frac{\sigma^2}{2} \tag{34}$$

or

$$3.21 - 2.71 = 0.5 = \frac{\sigma^2}{2} \tag{35}$$

Consequently, the values obtained from the integration and the intercept were:

$$D_1 = exp(2.71) = 15.0 \tag{36}$$

and

$$\sigma_1 = 1.0 \tag{37}$$

Regardless of whether the distribution is actually unimodal, the inversion procedure gives unambiguous answers.

The final step in the procedure is to test whether the distribution was actually unimodal. Using the inversion values of 15.0 and 1.0, numerical simulations covering efficiencies from 0.05 to 0.95 are carried out. These values indicated by open triangles are plotted in Figure 5. In this case, the triangles clearly overlap with the data (the solid curve) and the distribution is adequately represented by a single log-normal.

Were the simulated points to be significantly different from the data points, the hypothesis of a unimodal distribution would be rejected. The recommended procedure is to consider multimodal distributions with the restrictions:

$$\int_0^1 ln\overline{D}d\overline{Ef} = \Sigma C_j lnD_j \tag{38}$$

and

$$\lim_{\overline{Ef} \to 0} \overline{D}^n = \Sigma C_j D_j^n \exp(\overline{q}_j^2) \tag{39}$$

Frequently, inversion algorithms that are efficient for simulated measurements fail for real measurements. This is especially a problem in the case of diffusion batteries, where $n = 2/3$ and small errors in the penetration value are magnified by large uncertainties in the diffusion battery. In Table II, diffusion battery measurements with silver aerosols are presented. In this case, cylindrical channels of three different diameters were used. With a series of five flowrates, penetrations from 0.01 to 0.99 were obtained. For cylindrical pores, the fractional penetrations are given by:

$$Pt = 1 - 2.56 \phi^{2/3} + 1.2 \phi + 0.177 \phi^{4/3} \quad \text{for } \phi < 0.04 \tag{40}$$

$$Pt = 0.819 \exp(-3.65 \phi) + 0.097 \exp(-22.3 \phi) + 0.035 \exp(-57 \phi) \quad \text{for } \phi < 0.04 \tag{41}$$

$$\phi = \Pi N_t D^* L_t / Q$$

where N_t, L_t = number and length of channels
D^* = diffusion coefficient
Q = flowrate

For diffusion batteries, it is convenient to use the diffusion coefficient D^* rather than the diameter D as the size parameter. In Figure 6 the data

Table II. Silver Aerosols at 650°C

Flowrate (liter/min)	GCAF Diameter (μ)	Diffusion Coefficient (10^{-5} cm^2/sec)	Experimental Penetration	Simulated Penetration
1.0	10	4.6	0.005	0.006
2.2	10	7.2	0.03	0.03
4.8	10	10.9	0.10	0.10
7.3	10	11.8	0.15	0.16
1.0	25	10.6	0.13	0.13
2.2	25	13.3	0.30	0.28
4.8	25	19.4	0.42	0.45
7.3	25	21.1	0.51	0.54
1.0	50	20.0	0.75	0.75
4.8	50	19.7	0.91	0.90
7.3	50	24.0	0.92	0.93

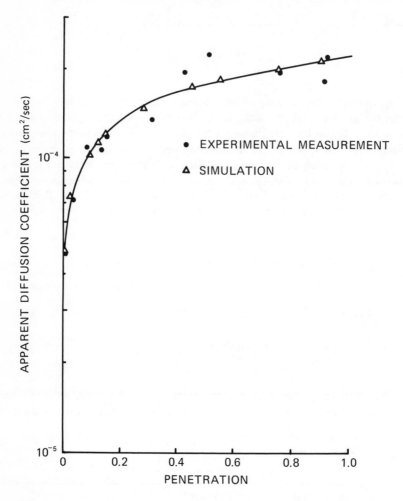

Figure 6. Apparent diffusion coefficient for silver aerosols measured with GCAF diffusion battery.

are plotted with the solid circles representing the experimental points. Numerical integration gave a value of -8.7 for $ln D_1^*$. From the intercept, a value for σ_1 of 1.3 was estimated. The solid curve represents simulated values based on these parameters. Based on this figure and the errors in the precision of the measurements as demonstrated by the scatter of the data points about the curve, it appears that a unimodal distribution adequately describes the data and that there appears no basis for testing more complicated distributions.

Two additional observations should be pointed out. The scatter in this experiment actually corresponds to a small difference between measured and simulated penetration (based on a log-normal distribution with the parameters above) and is quite small as can be seen from the last column in Table II. Secondly, one would expect that experimental uncertainties at low efficiencies (fractional penetrations near one) would be especially magnified when the apparent diffusion coefficients are calculated. With these data, this effect was not observed.

CONCLUSIONS

A method for determining the parameters of particle size distribution functions using the asymptotic behavior of the apparent diameter was developed. Assuming that the distribution function is described by a log-normal, the two parameters were determined as follows:

1. The log mean diameter was determined by a numerical integration of the logarithm of the apparent diameter as a function of efficiency. (This assumption is the Park-Gentry conjecture.)
2. The standard deviation was determined at the asymptotic condition of $ln\overline{D}$ as the efficiency approaches zero.

Although both of these conditions had been shown to work well with simulated data, they had not been previously tested with actual experiments nor has the method been examined theoretically. By assuming that the theoretical efficiency was given by a power law (exponential) relation, it was shown that:

1. The characteristic efficiency [the efficiency where the apparent diameter was equal to the mean diameter (log mean diameter) of the distribution] was a function of a single parameter q_1 proportional to the product of the standard deviation and the classifier exponent.
2. The shape of the apparent diameter-efficiency curve was strongly dependent on $\overline{q}_1$, even after the intercept was normalized (by dividing by $\overline{q}^2$). This effect was especially large at efficiencies near zero.
3. The parameters were obtained easily and unambiguously.
4. The Park-Gentry conjecture holds for an arbitrary sum of log-normal distributions; furthermore, only for $H(D) = lnD$ or $H(D) = 1$ is the equation:

$$\int_0^1 H(\overline{D})d\overline{Ef} = \int_0^\infty H(D)F(D)dD \qquad (42)$$

satisfied.

When the method was applied to actual measurements, it was shown that although there may be considerable scatter in the apparent diameters (or diffusion coefficients), since uncertainties in the efficiency (or penetration) measurement are magnified during transformation to apparent diameters, the method is efficient and non-ambiguous in providing estimates of the parameters.

ACKNOWLEDGMENTS

The work described here was partially supported by the state of Maryland Department of Natural Resources, under Grant P 67 80 04; the National Science Foundation, under Grant 80-11269; the Environmental Protection Agency under Grant R 80651801; and the University of Maryland Computer Center.

REFERENCES

Fuchs, N. A., J. B. Stechkina and V. I. Starosselskii (1962) *Brit. J. Appl. Phys.* 13:280.

Gormley, P. G., and M. Kennedy (1949) "Diffusion from Stream Flowing Through a Cylindrical Tube," *Proc. Roy. Irish Acad.* 52A:163-169.

King, W. E., J. W. Gentry and Y. O. Park (1980) "Determination of Distribution Functions from the Inversion of Fredholm Integral Equations," in *Multiphase Transport: Fundamentals, Reactor Safety, Applications, Vol. I*, T. Veziroglu, Ed. (Hemisphere Publications), pp. 323-340.

Maigne, J. P., P. Y. Turpin, G. Madelaine and J. Brichard (1974) "Nouvelle Methode de Determination de la Granulometric d'un Aerosol du Moyen d'une Batterie de Diffusion," *J. Aerosol Sci.* 5:339-355.

Megaw, W. J., and R. D. Wittin (1963) "Measurement of the Diffusion Coefficient of Homogeneous and Other Nuclei," *J. Rech. Atmos.* 1:113.

Mercer, T. T., and T. D. Greene (1974) "Interpretation of Diffusion Battery Data," *J. Aerosol Sci.* 5:251-256.

Nolan, P. J., and J. A. Scott (1963) "Observations on the Heterogeneity of Condensation Nuclei," *Proc. Royal Irish Acad.* 63A:35.

Nyström, E. J. (1930) "Über die Praktische Auflosung von Integralgleichungen mit Anwendungen auf Randwertaufgaben," *Acta Math.* 54:185-204.

Park, Y. O., W. E. King and J. W. Gentry (1980) "On the Inversion of Penetration Measurements to Determine Aerosol Product Size Distributions," *Ind. Eng. Chem. Prod. Res. Devel.* 19(2):151-157.

Perrin, M. L. (1980) "Etude de la Dynamique d'Aerosole fins Produits Artificiellement: Application á l'Atmosphere," PhD Dissertation, Paris.

Pollak, L. W., and A. L. Metnieks (1957) "On the Determination of the Diffusion Coefficient of Heterogeneous Aerosols by the Dynamic Method," *Geofis. Pura Appl.* 37:183-190.

Raabe, O. G. (1978) "A General Method for Fitting Size Distributions to Multicomponent Aerosol Data Using Weighted Least-Squares," *Environ. Sci. Technol.* 12:1162-1166.

Solderholm, S. C. (1979) "Analysis of Diffusion Battery Data," *J. Aerosol Sci.* 10:163-175.

CHAPTER 25

SAMPLING EFFICIENCY DETERMINATION OF AEROSOL SAMPLING INLETS

Klaus Willeke and Pål Å. Tufto*

Aerosol Research Laboratory
Department of Environmental Health
University of Cincinnati
Cincinnati, Ohio

ABSTRACT

New standards on particulate air sampling for the protection of human health will specify upper particle size cutoffs. This creates a demand for extensive testing of particulate sampling inlets. A wind tunnel has been designed and built that incorporates a new method for determining sampling efficiencies. The inlet under study is integrated into a modified optical single particle counter that records the aerosol concentration penetrated through the inlet. The penetrated aerosol concentration is thus measured dynamically and quickly for various particle sizes, sampling velocities, wind velocities and sampling angles.

By this technique, extensive measurements were performed of the overall sampling efficiency of a thin-walled sampling tube of 0.565 cm i.d. and 20 cm length. The sampling efficiency of the inlet tube was studied at wind velocities of 250–1000 cm/sec, inlet velocities of 125–1000 cm/sec, and angles from 0 to ±90°. The sampling efficiency was found to be significantly reduced when sampling was performed at an angle to the flow. When the sampling velocity in the inlet differed from the ambient wind velocity, the sampling efficiency was significantly increased or

---

*Present address: Division of Organization and Work Science, Norwegian Institute of Technology, University of Trondheim, Trondheim, Norway.

decreased. Differences in sampling efficiency were found for particles above 10 μm in diameter when the aerosol was sampled 15° upward vs 15° downward from the horizontal, downward sampling giving the higher sampling efficiency. For θ = 30–90°, the sampling efficiency was found to be approximately a function of Stokes number with the sampling ratio R = wind velocity to inlet velocity as a parameter. At θ = 90° the sampling efficiency was approximately a function of Stk·$\sqrt{R}$. About half or more of the particle deposition in the inlet occurred within the first 1 cm of the 20-cm-long inlet tube.

INTRODUCTION

When aerosols are sampled from ambient or industrial environments, air containing particles is drawn through an inlet opening to a filter or direct-reading instrument. It is essential that the sampled aerosol be representative of the aerosol upstream of the sampler, i.e., the aerosol concentration, size distribution, chemistry and other properties should be unchanged by the sampling process. If changes do occur, they should be known quantitatively and be independent of variations in conditions outside the sampler to the greatest extent possible.

Most of the theoretical and experimental studies have focused on the aspiration efficiency which considers sampling from the air environment to the face of the inlet [Agarwal and Liu 1980; Badzioch 1959; Belyaev and Levin 1972,1974; Bien and Corn 1971; Davies 1968; Davies and Subari 1979; Durham and Lundgren 1980; Fuchs 1975; Jayasekera and Davies 1980; Kaslow and Emrich 1974; Laktionov 1973; Levin 1957; Pattenden and Wiffen 1977; Rajendran 1979; Rüping 1968; Selden 1977; ter Kuile 1979; Vitols 1966; Watson 1954; Zebel 1978; Zenker 1971]. Studies considering the entire inlet have primarily evaluated specific inlet designs [Breslin and Stein 1975; Liu and Pui 1981; Lundgren and Calvert 1967; Ogden and Birkett 1978; Ogden and Wood 1975; Raynor 1970; Sehmel 1967,1970; Vincent and Gibson 1981; Wedding et al. 1967,1980].

The method described in this chapter is suitable for basic studies of entire aerosol sampling inlets as well as the calibration of simple or complex inlets. The performance of one specific inlet will be shown to demonstrate the dependence of overall sampling efficiency on air movements near the sampler.

DEFINITION OF SAMPLING EFFICIENCY

The sampling efficiency of an inlet for a specific particle size and flow condition may be defined as

$$E_s = c_s/c_0 \qquad (1)$$

where c_s = particle concentration measured after aspiration by and penetration through the inlet

c_0 = true particle concentration in the atmosphere

The true particle concentration may first be altered during aspiration to the face of the inlet, then by bounce of particles from the front edge of the inlet into the sampled airstream, and finally by losses to the inside wall while the aerosol is drawn through the length of the inlet. Overall sampling efficiency E_s thus may be represented by the product of three distinct efficiencies

$$E_s = E_a \cdot E_r \cdot E_t \qquad (2)$$

where E_a = aspiration efficiency

E_r = entry efficiency

E_t = transmission efficiency

Aspiration efficiency E_a is the ratio of the particle concentration at the face of the inlet to the particle concentration in the undisturbed environment. It is a function of the aerodynamic, inertial and gravitational forces acting on the particle. Entry efficiency E_r is the ratio of the particle concentration passing the inlet face to the particle concentration incident to that face. It is a function of the shape of the inlet's front edge, from which particles may rebound and be aspirated into the inlet. Particle rebound from the front edge of the inlet vs adhesion to that edge is generally not considered in trajectory calculations for the aspiration efficiency. Transmission efficiency E_t is the ratio of the particle concentration exiting from the inlet to the particle concentration just past the inlet face. It accounts for the particle losses to the inside wall by impaction, gravitational settling, and turbulent or laminar diffusion.

Most present standards on particulate air sampling from ambient or industrial environments require the measurement of the total suspended particulates (TSP), i.e., the total amount of airborne particles. This measurement does not specify an upper particle size limit. Each particulate sampler has its own upper particle size cutoff, which may vary with the speed and direction of the wind next to the sampler. To get more consistent results, new federal standards on particulate air sampling for the protection of human health will therefore specify upper particle size cutoffs [Hileman 1981; Miller et al. 1979].

Extensive performance testing of inlets will have to be undertaken to determine whether present and future aerosol samplers measure the airborne particle concentration according to the proposed size-specific standards. Also, more knowledge must be acquired on the sampling performance of simple inlets, so that the performance of more complex inlets may be predicted.

The sampling efficiency of a given inlet should be determined for solid and liquid particles of different sizes at different wind speeds and directions. In the traditional technique, tagged test aerosols are sampled through the inlet under study, and are deposited onto a filter. Generally, a conical expansion section is required ahead of the filter, to decelerate the flow to the appropriate filtration velocity. The quantity of aerosol deposits in the inlet and on the filter are then determined by fluoro- or radiometric techniques. Sampling efficiency determinations by such a technique are very time-consuming. Our method permits a much faster evaluation of the performance of a given inlet.

SAMPLING STRATEGY

Equation 2 identifies three component sampling efficiencies through subscripts, such as a for aspiration. To describe the measurement method used, we further subscript the efficiencies and corresponding particle concentrations by velocity ratio R and sampling angle θ. The ratio of wind velocity outside the sampler u_{wind} to the average air velocity in the inlet u_{inlet} defines the sampling kinetics through velocity ratio

$$R = u_{wind}/u_{inlet} \tag{3}$$

Sampling angle θ defines the sampling tube orientation through measurement of the angle between wind direction and inlet tube axis. The number concentration of sampled particles is thus

$$c_{s,R,\theta} = N/A \cdot u_{inlet} \tag{4}$$

where   N = count rate of particles registered by a particle counter downstream of the inlet
        A = cross-sectional area of the inlet

The true particle concentration in the wind tunnel is determined by sampling isokinetically with a thin-walled inlet for which the aspiration efficiency $E_{a,1,0}$ and the entry efficiency $E_{r,1,0}$ may be assumed equal to 100%, so that

$$E_{s,1,0} = E_{a,1,0} \cdot E_{r,1,0} \cdot E_{t,1,0} = E_{t,1,0} \tag{5}$$

The transmission efficiency $E_{t,1,0}$ is found by counting the particles penetrated through the inlet and adding to this the particles lost to the inner wall of the inlet as measured by a washoff technique.

Once the sampling efficiency is known for isokinetic flow, the sampling efficiency for all other kinetic conditions and tube orientations can be found by dynamic counting of the particle penetration through the inlet for these conditions and relating these counts to the one obtained for isokinetic flow. We can thus define a relative sampling efficiency $E_{rel,R,\theta}$

$$E_{rel,R,\theta} = c_{s,R,\theta}/c_{s,1,0} \tag{6}$$

which is measured as

$$E_{rel,R,\theta} = \frac{N_{s,R,\theta}/A \cdot u_{inlet}}{N_{s,1,0}/A \cdot u_{wind}} = \left(\frac{N_{s,R,\theta}}{N_{s,1,0}}\right) \cdot R \tag{7}$$

The total sampling efficiency of the inlet is then the product of Equations 5 and 7

$$E_{s,R,\theta} = E_{rel,R,\theta} \cdot E_{s,1,0} \tag{8}$$

In our new measurement method, the inlet under study is incorporated into an optical single particle counter, as described in the next section. The particle penetration is dynamically registered by the counter at all angles and velocity ratios of interest for all pertinent particle sizes (numerator of Equation 7). If the inlet is thin-walled, it is periodically operated isokinetically to give a reference count (denominator of Equation 7). If the inlet under study is thick-walled—for which the product of aspiration and entry efficiency may not be 100%—or if the inlet is of complex design, then the reference count is determined by a separate thin-walled inlet, and $E_{s,1,0}$ in Equation 8 is the reference efficiency of the thin-walled inlet.

To determine the true sampling efficiency (Equation 8), the relative sampling efficiency (Equation 7) is multiplied by the sampling efficiency at isokinetic conditions (Equation 5). The latter is found by operating the wind tunnel in the traditional way with uranine dye-tagged aerosols, and adding the number of fluorometrically determined particles deposited on the inner wall to the particle count registered by the dynamic counter [Tufto 1981]. Monodisperse particles are used for testing so that the deposited particle mass is readily converted to particle number for this calculation.

EXPERIMENTAL METHODS

We designed a special wind tunnel for sampling efficiency studies. The schematic representation of Figure 1 displays the essential elements. Laboratory air is drawn through four high-efficiency particulate air (HEPA) filters into the mixing chamber of the wind tunnel. The air is accelerated through a tapered section, which results in a constant-velocity profile throughout most of the 30- X 30-cm test section, as measured by a thermal anemometer probe (Model 430-3, Kurz Instruments, Carmel Valley, California).

Test aerosols are injected into the mixing chamber through an aerosol generator whose design is based on the vibrating-orifice principle [Berglund and Liu 1973]. In our design the ^{85}Kr charge neutralizer section is fixed, and the vibrating-orifice section is removable as a small unit. We abandoned the conventional use of liquid feeding to the generator by a syringe pump because we found variations in particle size as a function of liquid depletion in the plastic syringe. We replaced the syringe with a pressure-feed system as used in an earlier application of this generation principle [Raabe and Newton 1970]. In our design, the pressure in the liquid feed cylinder (emptied gas purifier, Alltech Assoc., Deerfield, Illinois) was kept constant through a high-precision pressure regulator (Moore Products, Spring House, Pennsylvania), and the liquid feed rate was calibrated by passing the liquid through a 0.2-ml pipet into which we injected an air bubble for measuring the liquid displacement rate. The aerosol was distributed in the mixing chamber by a large disc fan as used for window exhaust in homes. This resulted in well-mixed aerosol concentrations in the core of the test section. Conventional mixing baffles were found to eliminate too many particles above 20 μm in diameter. The turbulence level in the wind tunnel could be varied from 3 to 8% by adjustment in the speed of the mixing fan (measured by Series 1050 Anemometer System, TSI Inc., St. Paul, Minnesota). Turbulence variations in this range did not significantly affect test-inlet performance. Wedding et al. [1977] reported similar findings on a different inlet design.

The inlet under study samples from the aerosol flow in the test section of the wind tunnel. To sample dynamically, the inlet is integrated into the sensor of a modified optical single particle counter (Model 245/242, Royco Instruments Inc., Menlo Park, California). As seen in Figure 2, the sampling inlet is surrounded by clean sheath air, so that the aerosol leaves the inlet in or near the centerline of a converging circular channel

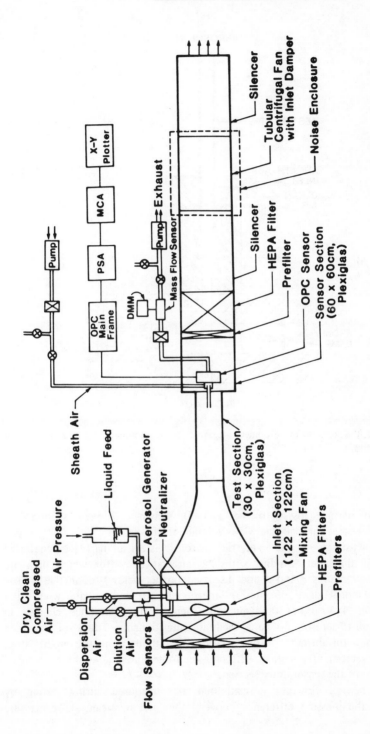

Figure 1. Wind tunnel for sampling efficiency studies. OPC = optical particle counter, PSA = pulse shaping amplifier, MCA = multichannel analyzer, DMM = digital multimeter.

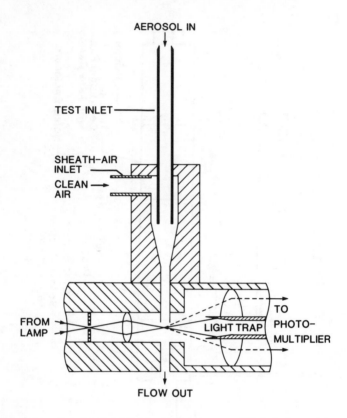

Figure 2.   Schematic of test inlet integrated into a modified single-particle optical counter.

without particle losses to the wall. On leaving the entry section, the aerosol passes through the view volume of the sensor. With no particles lost downstream of the inlet, the sensor registers all particles penetrated through the inlet. Studies with the circular inlet (for which data are shown in Figures 3 and 4 and discussed further below) found the particle to be independent of the volumetric sheath air flowrate when that rate was 4–40 times the volumetric aerosol flowrate. Such changes in sheath-air-to-aerosol flowrate facilitate the dynamic study of sampling efficiency for different velocity ratios R, without geometric modification of the sensor entry section. Different-sized inlets are studied by exchanging portions of the sensor entry section.

The aerosol flowrate is calibrated and measured during testing by noting the pressure difference between the inlet upstream static pressure

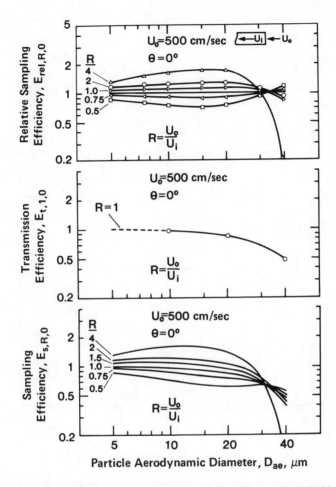

Figure 3. Sampling efficiency at isoaxial condition and wind velocity of 500 cm/sec. Thin-walled inlet tube, i.d. = 0.565 cm, o.d. = 0.635 cm (0.25 in.), L = 20 cm, $u_o = u_{wind}$, $u_i = u_{inlet}$.

and a static pressure tap near the inlet exit. Most measurements were made with the face of the inlet protruding about 15 cm into the 30- X 30-cm test section and with the sensor secured in a 60- X 60-cm downstream section. The airflow deflects around the body of the sensor and does not affect the flow pattern at the face of the inlet as verified by thermal anemometer measurements of the flow velocity near the inlet for the 250- to 1000-cm/sec flow velocity range studied. The

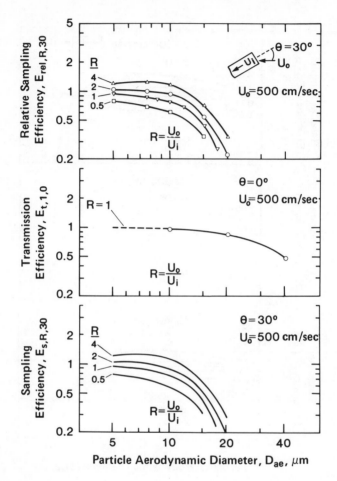

Figure 4.   Sampling efficiency at $\theta = +30°$ and wind velocity of 500 cm/s. Thin-walled inlet tube, i.d. = 0.565 cm, o.d. = 0.635 cm (0.25 in.), L = 20 cm, $u_o = u_{wind}$, $u_i = u_{inlet}$.

angle of the inlet can be varied from parallel ($\theta = 0°$) to perpendicular ($\theta = \pm90°$) to the wind tunnel flow. When sampling at ±60 to $\pm90°$ to the flow, the sensor is located external to the test section.

The optics of the particle sensor are modified to give a larger view volume to accommodate flowrates up to 75 liter/min through the sensor. Many low- and medium-flow inlets can thus be integrated into the optical particle counter. The larger view volume has a greater range in illumination intensities in it, so that monodisperse particles are recorded over a wider range of channels (Model 60 Multi-Channel Analyzer, Nuclear Data Inc., Schaumburg, Illinois). The sheath airflow is supplied by an air mover, and the combined flow is withdrawn from the sensor by a pump outside the wind tunnel.

Downstream of the sensor section the aerosols are removed by in-line filters. The air mover for the wind tunnel (Tubular Acousta Foil, Size T189, New York Blower Co., Chicago, Illinois) is inline with two duct silencers (Model 3SR24-24, Environmental Elements Corp., Dallas, Texas) that reduce the aerodynamic flow noise. The air mover is also surrounded by a ventilated enclosure, which reduces the noise radiated by the body of the fan (see Figure 1).

EXAMPLES OF SAMPLING EFFICIENCY DETERMINATION

The new method was used to evaluate the sampling efficiency of a thin-walled inlet tube of 20 cm length, 0.565 cm inner diameter and 0.635 cm (0.25 in.) outside diameter. To exemplify our method, the measured sampling efficiencies at a wind velocity of 500 cm/sec are shown for sampling parallel to the flow direction ($\theta = 0°$, Figure 3) and 30° downward from the flow direction ($\theta = +30°$, Figure 4). Oleic acid particles of 5–40 μm diameter were used as test aerosols. The upper graphs show the relative sampling efficiencies of the inlet, as calculated from the optical particle count of the penetrated particle concentration at the angle and velocity ratio under study and at isoaxial, isokinetic conditions (Equation 7). The middle graphs give the transmission efficiency for isokinetic sampling at the stated wind velocity. The transmission efficiency is obtained by relating the particles lost in the inlet to the particles penetrated through the inlet. For the thin-walled inlet studied, the sampling efficiency at isokinetic conditions $E_{s,1,0}$ equals the transmission efficiency $E_{t,1,0}$ (Equation 5). The overall sampling efficiency $E_{s,R,\theta}$ (Equation 8) is the product of the two efficiencies shown, and is displayed in the lower graphs of Figures 3 and 4.

Relative sampling efficiency data for variations in angle and velocity ratio are obtained quickly. Since only a few of the time-consuming washoff experiments need to be performed, the sampling efficiency of a given inlet can be determined rapidly for a wide range of conditions.

RESULTS

Experiments with the sampling inlet were performed with different inlet velocities and sampling angles for a set of particle sizes at wind velocities of 250, 500 and 1000 cm/sec. The data of size-dependent sampling efficiencies will be presented for different angles with a fixed inlet velocity, wind velocity or ratio of the two velocities.

Figure 5 presents the sampling efficiency for fixed inlet velocities of 250 and 1000 cm/sec. Since industrial and environmental samplers generally

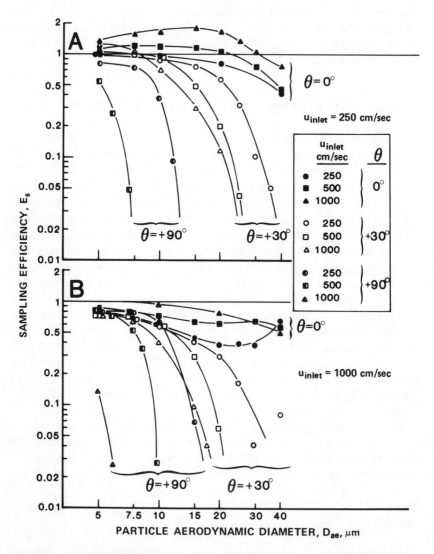

Figure 5. Sampling efficiency of a thin-walled inlet tube at constant inlet velocity. L = 20 cm, i.d. = 0.565 cm, o.d. = 0.635 cm (0.25 in.), A: inlet velocity = 250 cm/sec, B: inlet velocity = 1000 cm/sec.

draw air at a fixed volumetric flowrate, this type of presentation allows the practitioner to evaluate the sampler's performance in the most direct manner. For instance, the closed-face filter cassette that is conventionally used by the industrial hygienist has an inlet velocity of 265 cm/sec when operated at 2 liter/min through its 4-mm-diameter opening. The Climet Model 208 optical single-particle counter has a similar-sized opening and an inlet velocity of about 1000 cm/sec.

Figure 5 shows that, for $\theta = 0°$, the sampling efficiency increases with increasing wind velocity as expected from differences in aspiration efficiency. At $\theta = 30°$, for particles above 10 μm, the sampling efficiency decreases with increasing wind velocity because particles increasingly impact onto the inner wall just past the inlet face as the wind velocity increases. At $\theta = 90°$, an increase in wind velocity results in a decreased sampling efficiency for all particle sizes due to decreased aspiration (particles overshooting the inlet) and increased impaction as the wind velocity increases. Figure 5 also shows that for the wind velocities tested, the sampling efficiency at an inlet velocity of 250 cm/sec is higher at $\theta = 0°$, and lower at $\theta = \pm90°$ than the sampling efficiency for the same angles at an inlet velocity of 1000 cm/sec.

Studies of the aspiration efficiency, which is concerned with the sampling of particles from the air environment to the face of the inlet, have shown that the aspiration efficiency can be represented in generalized form as a function of Stokes number Stk

$$Stk = \frac{\tau \cdot u_{wind}}{D} \qquad (9)$$

where u_{wind} = wind velocity
   D = inner diameter of the inlet
   τ = particle relaxation time [Belyaev and Levin 1972,1974; Davies 1968; Davies and Subari 1979; Durham and Lundgren 1980]

The particle relaxation time is a second power function of the physical particle diameter, and includes the particle density and viscosity of the suspending gas medium. Stk is then an impaction parameter and represents particle size in a nondimensional form. In consideration of the aspiration efficiency, the air velocities are also made nondimensional through the velocity ratio R defined previously in Equation 3.

In an actual inlet, the true particle concentration in the air environment may be modified not only by aspiration to the face of the inlet, but also by particle rebound from the leading edge of the inlet, impaction onto the inner wall of the inlet, turbulent deposition, gravitational settling and, when submicrometer particles are present, diffusional deposition on the inner wall. Figure 6 plots the overall sampling efficiency of the given inlet as a function of Stk for velocity ratios of 1 and 2. The particle

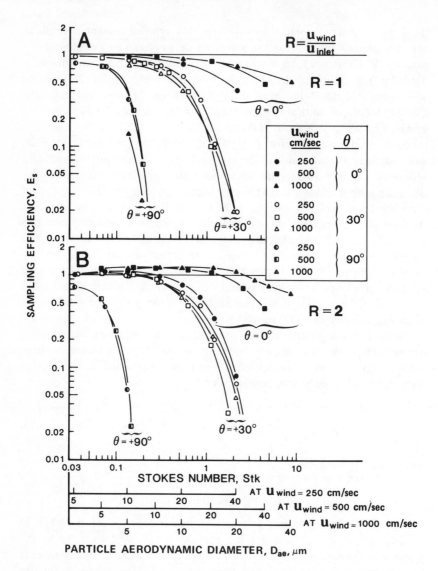

Figure 6. Sampling efficiency of a thin-walled inlet tube at constant sampling ratio. L = 20 cm, i.d. = 0.565 cm, o.d. = 0.635 cm (0.25 in.), A: R = 1, B: R = 2.

aerodynamic diameters are also shown for the three different wind velocities tested. Our data, of which some are presented in Figure 6, show that the sampling efficiency is approximately a function of Stk at sampling angles of 30, 60 and 90° with velocity ratio R as a parameter. However, under isoaxial sampling conditions ($\theta = 0°$) the sampling efficiency is not a function of Stk. Impaction is thus not the dominant particle removal mechanism in the latter case. As further shown below, gravitational settling in the inlet tube appears to be the primary removal mechanism of large particles when sampling isoaxially, whereas impaction onto the inner wall just past the face of the inlet appears to be the primary particle removal mechanism when sampling at an angle larger than about 30°.

An effect of gravitational settling of particles in the inlet is interpreted through Figure 7, which shows the measured sampling efficiency for upward vs downward sampling at a 15° angle with the wind velocity = 500 cm/sec. As seen, upward sampling has a lower sampling efficiency than downward sampling for particle aerodynamic diameters greater than about 10 μm. This was also found at wind velocities of 250 and 1000 cm/sec. The difference between upward and downward sampling decreases as the wind velocity increases [Tufto 1981]. Apparently, the relative effect of the gravitational settling velocity on the deposition process in the front end of the tube is reduced as the wind velocity is increased. This will be further explained in the discussion section. In all cases, the difference between upward and downward sampling was found to increase with increasing particle size. A difference in sampling efficiency between upward and downward sampling was also found at θ = 30° for particle sizes above 10 μm, although the effect was not as pronounced as for θ = 15°. At θ = 60°, the sampling efficiency for upward sampling was found to be only slightly lower than for downward sampling. At θ = 90°, no difference was found between upward and downward sampling.

When the sampling angle is increased from 0 to 90°, the sampling efficiency decreases. Figure 8 shows the sampling efficiency at a wind velocity of 500 cm/sec for different sampling angles with the velocity ratio R as a parameter. At isoaxial sampling (θ = 0°) the sampling-efficiency curve for the lowest inlet velocity of 125 cm/sec (R = 4) is seen to drop off so much at the larger particle sizes that it crosses over the other curves. All the θ = 0° curves are seen to cross each other at a particle size of about 30 μm. The sampling efficiency is thus higher for low than for high inlet velocities when small particles are sampled, but lower for low than for high inlet velocities when large particles are sampled. When the inlet velocity is decreased, the number concentration of particles aspirated by the inlet increases. When the particles are sufficiently large, however, gravity removes a significant portion of the particles along the length of the inlet tube, with the amount depending on the flow time through the inlet tube. A decrease in inlet velocity thus decreases the sampling efficiency and causes the crossover of efficiency curves at 30 μm.

Comparisons of the isoaxial experiments at the wind velocities tested (250, 500 and 1000 cm/sec) have shown that the sampling efficiencies of the given inlet are approximately a function of Stk and R for the

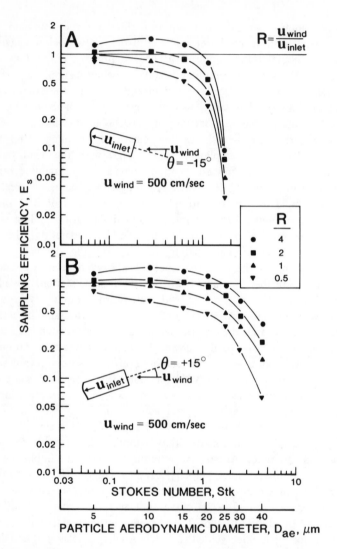

Figure 7. Sampling efficiency of a thin-walled inlet tube for downward vs upward sampling. L = 20 cm, i.d. = 0.565 cm, o.d. = 0.635 cm (0.25 in.), A: upward sampling at $\theta = -15°$, B: downward sampling at $\theta = +15°$.

smaller particle sizes (Stk < 0.15). The crossover points of the data lines occur at a particle size of about 20–40 μm and are not a function of Stokes number.

By increasing the angle to 15 or 30°, no crossover was found at wind velocities of 500 or 1000 cm/sec. The sampling efficiencies at an angle

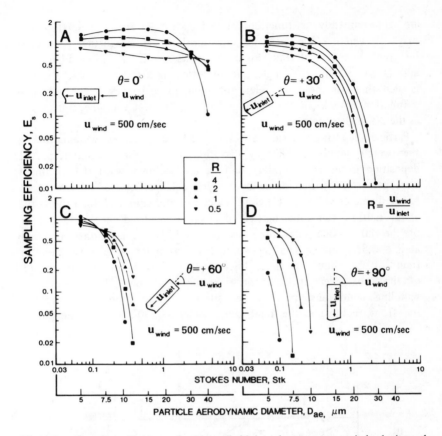

Figure 8.   Sampling efficiency of a thin-walled inlet tube at constant wind velocity and different angles. L = 20 cm, i.d. = 0.565 cm, o.d. = 0.635 cm (0.25 in.), A: $\theta = 0°$, B: $\theta = +30°$, C: $\theta = +60°$, D: $\theta = +90°$.

of 30° and a wind velocity of 500 cm/sec are shown in Figure 8B. At a wind velocity of 250 cm/sec, some of the 15 and 30° data lines were found to cross each other.

At an angle of 60°, crossover occurs at about Stk = 0.1 for wind velocities of 250 and 500 cm/sec. The results for a wind velocity of 500 cm/sec are shown in Figure 8C. At a wind velocity of 1000 cm/sec, crossover can be extrapolated to Stk ≈ 0.1, below the 5-μm particle size tested.

As shown in Figure 8D, no crossover occurs within the experimental range for an angle of 90° at a wind velocity of 500 cm/sec. Note, however, that high R values result in lower efficiencies at $\theta = +90°$, whereas they result in higher efficiencies at $\theta = +30°$ when compared to low R values.

Figure 6 showed that the sampling efficiencies at $\theta = +30$ and $+90°$

are approximately a function of Stk with R as a parameter. One may ask whether the efficiency relationships can be further generalized by combining R with Stk. Figure 9 shows the results when Figures 8B and D for 30 and 90° angles are replotted as a function of $Stk\sqrt{R}$. It is seen that for the given inlet, the sampling efficiency is approximately a function of $Stk\sqrt{R}$ for an angle of 90°, but not for other angles such as the 30° angle shown.

Particle deposition inside the sampling tube causes a reduction in the overall sampling efficiency. To find out the distribution of particle deposition inside the sampling tube, a few sectional washoff experiments were performed with uranine-tagged oleic acid particles of 20 μm aerodynamic diameter sampled at a wind velocity of 500 cm/sec. Table I shows that about half of the total particle deposition occurred in the first 1 cm of the 20-cm-long inlet tube, when sampling was performed isoaxially at R = 0.5, 1 and 2. The deposition per centimeter of tube was therefore found to be about 20 times higher in the first 1 cm than in the rest of the tube, even at isokinetic sampling conditions. Under nonisoaxial sampling conditions, particle deposition is even more pronounced in the front section of the inlet tube. At θ = +30°, 76% of the particle

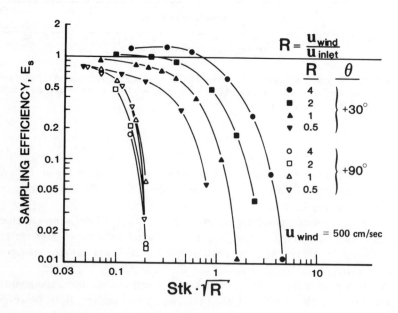

Figure 9. Sampling efficiency dependence on $Stk\sqrt{R}$. L = 20 cm, i.d. = 0.565 cm, o.d. = 0.635 cm (0.25 in.).

Table I. Sectional Distribution of Particle Deposition in the Sampling Tube[a]

Angle θ	u_{inlet} (cm/sec)	R, u_{wind}/u_{inlet}	Fractional Deposition per cm of Tube (%)			
			0–1 cm	1–7.3 cm	7.3–13.7 cm	13.7–20 cm
0°	250	2	60.2	3.0	1.8	1.6
0°	500	1	46.6	2.7	3.6	2.2
0°	1000	0.5	50.8	2.1	2.9	2.8
+30°	500	1	75.8	3.8	0.02	0.02

[a]Oleic acid test particles of 20 μm aerodynamic diameter at wind velocity = 500 cm/sec.

deposition was found in the first 1 cm of the inlet tube, and 99.6% in the first 4 cm of the tube.

The amount of particle deposition in the inlet tube determines the transmission efficiency, which we are defining as the ratio of the particle concentration exiting from the inlet tube to the particle concentration at the inlet face of the tube. The latter is found by combining the particle deposition data with simultaneous particle penetration measurements obtained from the optical single-particle counter. Table II shows percent transmission efficiencies for the four conditions of Table I and for two additional isokinetic cases. The second row of Table II shows that the transmission efficiency of the inlet tube is lower for subisokinetic sampling ($u_{inlet} < u_{wind}$) than for superisokinetic sampling of 20-μm particles. The transmission efficiency for the nonisoaxial condition is low (21.1%).

One may also calculate the aspiration efficiency by dividing the overall sampling efficiency of the thin-walled inlet tube by its transmission efficiency. For the isoaxial cases within inlet velocities of 250 and 1000 cm/sec at a wind velocity of 500 cm/sec, the aspiration efficiencies

Table II. Percent Transmission Efficiency[a]

Angle θ	u_{wind} (cm/sec)	u_{inlet} (cm/sec)		
		250	500	1000
0°	250	81.1		
0°	500	62.8	84.6	94.1
0°	1000			76.6
+30°	500		21.1	

[a]Oleic acid test particles of 20 μm aerodynamic diameter.

for 20-μm particles are 170 and 66%, respectively, which are in good agreement with the theoretical predictions of Belyaev and Levin [1972, 1974]. For the 30° case in Table II, with the inlet velocity equal to the wind velocity, the calculated aspiration efficiency is 51%, which is lower than the one reported by Durham and Lundgren [1980].

DISCUSSION

The extensive sampling efficiency data we have obtained for one specific inlet provide the basis for our attempt to identify the dominant particle removal mechanisms in the inlet. The tests were performed with liquid particles. Solid particles may rebound from the inner wall of the inlet tube and thus result in higher sampling efficiencies.

The data presented may be used for estimating the sampling efficiencies of inlets in use, such as the opening of the closed-face filter cassette and the inlet tube of the optical single-particle counter mentioned earlier. However, if there is a difference in tube length, caution should be exercised, especially in using the isoaxial sampling efficiency results. Although most of the particle deposition in our 20-cm-long inlet was found to occur close to the inlet face (Table I), there is a significant dependence on tube length in our isoaxial data. For the nonisoaxial data, however, the dependence on tube length is insignificant as long as the tube is above a certain minimum length of a few centimeters. Most inlets in use are blunt or thick-walled and may therefore have somewhat different sampling efficiencies than we have obtained with our thin-walled inlet.

Figure 10 displays the dominant particle removal mechanisms as we interpret them from our data. For isoaxial sampling ($\theta = 0°$), particle aspiration to the inlet face appears to affect the sampling most for small particles, whereas gravitational sedimentation in the horizontal inlet tube appears to be the dominant removal mechanism for large particles. Aspiration-dominated sampling results in undersampling (i.e., lower sampling efficiencies) when the inlet velocity is higher than the wind velocity (see Figures 8A and 10A). Aspiration efficiencies can be plotted as a function of Stk with R as a parameter. Figure 6 also shows that the overall sampling efficiency for $\theta = 0°$ can be plotted in that manner for small particles, but not for large ones. For large particles, the sampling efficiency appears to be more closely described by the gravitational deposition parameter Z

$$Z \propto \frac{v_s \cdot L}{u_{inlet} \cdot D} \tag{10}$$

where   L = length of the inlet tube
        v_s = gravitational settling velocity [Schwendiman et al. 1975]

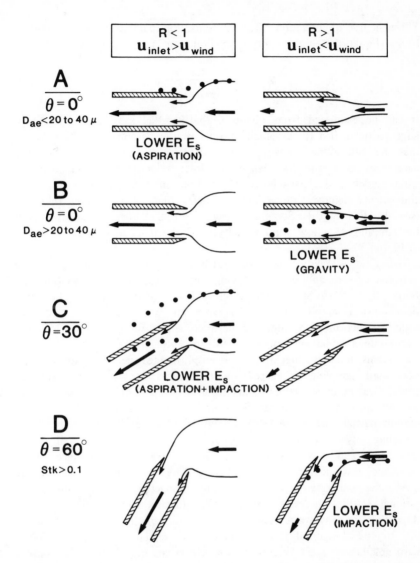

Figure 10. Dominant mechanisms for particle losses when sampling through a thin-walled inlet tube.

The latter is a function of particle relaxation time τ and gravitational acceleration g

$$v_S = \tau \cdot g \tag{11}$$

Through use of Equation 11, the deposition parameter can be related to Stk (Equation 9) and velocity ratio (Equation 3) as:

$$Z \propto g \cdot L \cdot R \cdot \frac{Stk}{(u_{wind})^2} \qquad (12)$$

It can be estimated from Figure 6 that isoaxial sampling of relatively large particles at a fixed velocity ratio is better described by $Stk/(u_{wind})^2$ than by Stk alone. Gravitational settling causes the crossover of the data lines in Figure 8A. When the inlet velocity is smaller than the wind velocity, the particle residence time in the inlet tube is longer and gravity removes more large particles than at high inlet velocities (see also Figure 10B). Some crossover of the data lines was also observed for sampling at a low wind velocity of 250 cm/sec and moderate angles of 15 and 30°.

The effect of gravity is also apparent in the sampling efficiency difference between upward and downward sampling (see Figure 7). It was generally found that downward sampling resulted in higher sampling efficiencies than upward sampling for particle sizes above 10 μm. The effect was found to be most pronounced between angles of +15° and −15°, and was not detected between angles of +90° and −90°.

It seems unlikely that the observed difference between upward and downward sampling is caused by differences in aspiration efficiency alone. Gravitational settling along the inside wall of the inlet tube should not be much different either. We propose that the differences are mainly due to particle impaction onto the inside wall near the inlet face. When sampling downward, impaction occurs on the upper inside wall of the inlet tube and gravity pulls away from the impaction surface. When sampling upward, impaction occurs on the lower inside wall of the inlet tube and gravity pulls toward the impaction surface. Thus, gravitational settling increases impaction in upward sampling but decreases impaction in downward sampling. As the sampling angle approaches 90° to the horizontal, the component of gravitational settling velocity perpendicular to the tube wall becomes zero and no difference is observed between upward and downward sampling. However, since the magnitude of sampling efficiency decreases with increase in angle, and no particles were sampled above 10 μm for a wind velocity of 500 cm/sec and angles of ±90°, gravity becomes less important with increase in angle.

As the orientation of the inlet is tilted away from the wind vector, particle loss due to lack of aspiration increases, but impaction onto the inside wall near the inlet face also increases. The sectional washoff experiments shown in Table I indicate that most of the particles lost inside the tube are deposited just past the inlet face. At θ = 30° both effects appear to be important, as illustrated in Figure 10C. Since aspiration and impaction are both Stk-dependent, the curves of overall sampling efficiency are approximately functions of Stk for fixed velocity ratios, as seen in Figure 6. At θ = +30° the sampling efficiency of the example in Figure 8B is reduced to zero between 20 and 30 μm without any crossover. At θ = +60°, Figures 8C and 10D, no particles above 15 μm are sampled, but crossover occurs at a particle size of 6–7 μm where Stk ≈ 0.1. For θ = 60° at Stk > 0.1, where impaction seems to dominate, losses are greatest for conditions of $u_{inlet} < u_{wind}$ for which the particle trajectory has the least opportunity for turning into the inlet, as illustrated in Figure 10D. Sampling at θ = 90° appears to be similar to the case illustrated for θ = 60° in Figure 10D, although particles overshooting the inlet cause low aspiration efficiency, which may significantly affect sampling efficiency.

For sampling at 90°, our data plot well as a function of $Stk \cdot \sqrt{R}$, as shown in Figure 9. The data of Pattenden and Wiffen [1977] with an inlet at 90° also plot well in this manner [Davies and Subari 1979]. The point sink formula of Levin [1957], adapted to a finite-size sampling inlet, has been used to explain this dependence [Davies and Subari 1979].

In the study by Laktionov [1973] a polydisperse oil aerosol was sampled at an angle of 90° for a range of Stokes numbers from 0.003 to 0.2 and a range of velocity ratios from 1.25 to 6.25. The following empirical equation was fitted to the data of this study:

$$E_s = 1 - 3Stk^{1/\sqrt{R}} \tag{13}$$

For $1 \leqslant R \leqslant 4$, Equation 13 agrees well with our 90° results at Stk < 0.1. For R = 0.5 and for $1 \leqslant R \leqslant 4$ at Stk > 0.1, our sampling efficiencies are lower than those predicted by Equation 13.

ACKNOWLEDGMENTS

This research was supported by the U.S. National Institute for Occupational Safety and Health under Grant No. OH 00774; by the Norwegian Institute of Technology; by the Royal Norwegian Council for Scientific and Industrial Research; by funds from Center Grant USPHS ES 00159. We wish to thank A. Fodor and J. Svetlik for their technical assistance.

REFERENCES

Agarwal, J. K., and B. Y. H. Liu (1980) "A Criterion for Accurate Aerosol Sampling in Calm Air," *Am. Ind. Hygiene Assoc. J.* 41:191.

Badzioch, S. (1959) "Collection of Gas-Borne Dust Particles by Means of an Aspirated Sampling Nozzle," *Brit. J. Appl. Phys.* 10:26.

Belyaev, S. P., and L. M. Levin (1972) "Investigation of Aerosol Aspiration by Photographing Particle Tracks under Flash Illumination," *J. Aerosol Sci.* 3:127.

Belyaev, S. P., and L. M. Levin (1974) "Techniques for Collection of Representative Aerosol Samples," *J. Aerosol Sci.* 5:325.

Bergland, R. N., and B. Y. H. Liu (1973) "Generation of Monodisperse Aerosol Standards," *Environ. Sci. Technol.* 7:147.

Bien, C. T., and M. Corn (1971) "Adherence of Inlet Conditions for Selected Aerosol Sampling Instruments to Suggested Criteria," *Am. Ind. Hygiene Assoc.* 32:453.

Breslin, J. A., and R. L. Stein (1975) "Efficiency of Dust Sampling Inlets in Calm Air," *Am. Ind. Hygiene Assoc. J.* 36:576.

Davies, C. N. (1968) "The Entry of Aerosols into Sampling Tubes and Heads," *Staub-Reinhalt. Luft* (English) 28(6):1.

Davies, C. N., and M. Subari (1979) "Inertia Effects in Sampling Aerosols," in *Proceedings of the Symposium on Advances in Particle Sampling and Measurement*, EPA Report 600/7-79-065, pp. 1-29.

Durham, M. D., and D. A. Lundgren (1980) "Evaluation of Aerosol Aspiration Efficiency as a Function of Stokes Number, Velocity Ratio and Nozzle Angle," *J. Aerosol Sci.* 11:179.

Fuchs, N. A. (1975) "Sampling of Aerosols," *Atmos. Environ.* 9:697.

Hileman, B. (1981) "Particulate Matter: The Inhalable Variety," *Environ. Sci. Technol.* 15:983.

Jayasekera, P. N., and C. N. Davies (1980) "Aspiration below Wind Velocity of Aerosols with Sharp Edged Nozzles Facing the Wind," *J. Aerosol Sci.* 11:535.

Kaslow, D. E., and R. J. Emrich (1974) "Particle Sampling Efficiencies for an Aspirating Blunt Thick-Walled Tube in Calm Air," Technical Report No. 25, Department of Physics, Lehigh University, Bethlehem, Pennsylvania.

Laktionov, A. B. (1973) "Aspiration of an Aerosol into a Vertical Tube from a Flow Transverse to It," (Translation from Russian), AD-760 947, Foreign Technology Division, Wright-Patterson Air Force Base, Ohio.

Levin, L. M. (1957) "On the Sampling of Aerosols," *Bull. Acad. Sci. USSR, Geophys. Ser. Part II* 7:87.

Liu, B. Y. H., and D. Y. H. Pui (1981) "Aerosol Sampling Inlets and Inhalable Particles," *Atmos. Environ.* 15:589.

Lundgren, D. A., and S. Calvert (1967) "Aerosol Sampling with a Side Port Probe," *Am. Ind. Hygiene Assoc. J.* 28:208.

Miller, F. J., D. E. Gardner, J. A. Graham, R. E. Lee, W. E. Wilson and J. D. Bachmann (1979) "Size Considerations for Establishing a Standard for Inhalable Particles," *J. Air Poll. Control Assoc.* 29:610.

Ogden, T. L., and J. L. Birkett (1978) "An Inhalable-Dust Sampler for Measuring the Hazard from Total Airborne Particulate," *Ann. Occup. Hyg.* 21:41.

Ogden, T. L., and J. D. Wood (1975) "Effects of Wind on the Dust and Benzene-Soluble Matter Captured by a Small Sampler," *Ann. Occup. Hygiene* 17:187.

Pattenden, N. J., and R. D. Wiffen (1977) "The Particle Size Dependence of the Collection Efficiency of an Environmental Aerosol Sampler," *Atmos. Environ.* 11:677.

Raabe, O. G., and G. J. Newton (1970) "Development of Techniques for Generating Monodisperse Aerosols with the Fulwyler Droplet Generator," Report LF-43, Lovelace Foundation, Albuquerque, New Mexico, pp. 13-17.

Rajendran, N. (1979) "Theoretical Investigation of Inlet Characteristics for Personal Aerosol Samplers," Contract No. 210-78-0092, IIT Research Institute, Chicago, IL.

Raynor, G. S. (1970) "Variation in Entrance Efficiency of a Filter Sampler with Air Speed, Flow Rate, Angle and Particle Size," *Am. Ind. Hygiene Assoc. J.* 31:294.

Rüping, G. (1968) "The Importance of Isokinetic Suction in Dust Flow Measurements by Means of Sampling Probes," *Staub-Reinhalt. Luft* (English) 28(4):1.

Schwendiman, L. C., G. E. Stegen and J. A. Glissmeyer (1975) "Methods and Aids for Assessing Particle Losses in Sampling Lines," Report BNWL-SA-5138, Battelle Pacific Northwest Laboratories, Richland, WA.

Sehmel, G. A. (1967) "Errors in the Subisokinetic Sampling of an Air Stream," *Ann. Occup. Hygiene* 10:72.

Sehmel, G. A. (1970) "Particulate Sampling Bias Introduced by Anisokinetic Sampling and Deposition Within the Sampling Line," *Am. Ind. Hygiene Assoc. J.* 31:758-771.

Selden, M. G. (1977) "Estimates of Errors in Anisokinetic Sampling of Particulate Matter," *J. Air Poll. Control Assoc.* 27:235.

ter Kuile, W. M. (1979) "Comparable Dust Sampling at a Workplace," Report F 1699, Research Institute for Environmental Hygiene, Delft, The Netherlands.

Tufto, P. Å. (1981) "Sampling Efficiencies of Particulate Sampling Inlets," PhD Thesis, Department of Environmental Health, University of Cincinnati.

Tufto, P. Å., and K. Willeke (1982) "Dependence of Particulate Sampling Efficiency on Inlet Orientation and Flow Velocities," *Am. Ind. Hyg. Assoc. J.*

Tufto, P. Å., and K. Willeke (in press) "Dynamic Evaluation of Aerosol Sampling Inlets," *Environ. Sci. Technol.*

Vincent, J. H., and H. Gibson (1981) "Sampling Errors in Blunt Dust Samplers Arising from External Wall Loss Effects," *Atmos. Environ.* 15:703.

Vitols, V. (1966) "Theoretical Limits of Error Due to Anisokinetic Sampling of Particulate Matter," *J. Air Poll. Control Assoc.* 16:79.

Watson, H. H. (1954) "Errors Due to Anisokinetic Sampling of Aerosols," *Am. Ind. Hygiene Assoc. Quart.* 15:21.

Wedding, J. B., A. R. McFarland and J. E. Cermak (1977) "Large Particle Collection Characteristics of Ambient Aerosol Samplers," *Environ. Sci. Technol.* 11:387.

Wedding, J. B., M. Weigand, W. John and S. Wall (1980) "Sampling Effectiveness of the Inlet to the Dichotomous Sampler," *Environ. Sci. Technol.* 14:1367.

Zebel, G. (1978) "Some Problems in the Sampling of Aerosols," in *Recent Developments in Aerosol Science* D. T. Shaw, Ed. (New York: John Wiley & Sons, Inc.).

Zenker, P. (1971) "Investigations into the Problem of Sampling from a Partial Flow with Different Flow Velocities for the Determination of the Dust Content in Flowing Gases," *Staub-Reinhalt. Luft* (English) 31(6):30.

CHAPTER 26

SAMPLING ARTIFACTS IN THE BREATHING ZONE

B. S. Cohen, N. H. Harley, C. A. Martinelli and M. Lippmann
Institute of Environmental Medicine
New York University Medical Center
New York, New York

ABSTRACT

Factors that account for the discrepancy between estimates of beryllium inhalation exposure from different air sampling methods have been investigated. Discrepancies may result from artifacts of the sampling process or the variability of aerosol concentration within the breathing zone. Significant potential sources of error are: bias due to inlet effects, electrostatic effects due to fields developed by the polystyrene cassette personal monitors, filter efficiency and self-dilution by the samplers. The Be particle size distribution was measured with a new adaptation of the multicyclone sampler at a Be metal alloy casting operation. The particles present at the furnace area usually are in the size range where the efficiency of the personal monitor is near 100%. During certain operations, large particles are present and may be oversampled. Experiments in a specially designed dust room show that air cleaning occurs; however, field experiments indicate that the effect is not large. A portable backpack incorporating three continuous reading nephelometric aerosol monitors was developed and used for field monitoring of the aerosol concentrations in the breathing zone of a worker. The ratios of dust concentrations measured at different body positions were very variable. Although it is not yet possible to quantitatively apportion the differences, together these factors account for the discrepancy in exposure estimates.

347

INTRODUCTION

Comparative sampling to determine the inhalation exposure of workers to airborne beryllium by personal monitoring and the time weighting of representative breathing zone samples generally results in significantly higher exposure estimates from the personal monitors (PM) [Donaldson and Stringer 1980]. Reasons for the lack of agreement must be understood to estimate the true inhalation exposure. Discrepancies may result from artifacts of the sampling process or from variability of aerosol concentration within the breathing zone. This chapter describes our investigation of these factors in laboratory studies and in field studies at the same Be refinery studied by Donaldson and Stringer [1980]. Significant potential sources of error are: bias due to inlet effects, electrostatic effects due to fields developed by the polystyrene cassette PM and self-dilution by the samplers, by air cleaning or displacement of contaminated air.

The PM are three-piece polystyrene cassettes, holding 37-mm-diameter filters. Aerosol is drawn through a 4-mm-diameter circular inlet onto Type AA membrane filters (Millipore Corp.). The PM operates for a full 8-hour shift at a flowrate of 0.002 m^3/min. The inlet velocity is 2.6 m/sec. General air and breathing zone samples to calculate time-weighted averages (TWA) are taken with a "mini-Hi-Vol" (MHV) sampler. Dust is collected onto a 10.5-cm Whatman 41 filter paper at a flowrate of 0.25 m^3/min. The face velocity at the filter, with allowance for the filter surface covered by the retaining ring, is 0.56 m/sec.

INLET BIAS

According to the criteria of Davies [1968], no sampling bias is expected to result from the larger-aperture MHV sampler in calm air with the filter in a vertical plane. He recommended that for an aerodynamic diameter of 10 μm, wind speed should not exceed 0.1 m/sec to obtain an efficient sample. For the PM, the small sampling orifice is expected to discriminate against larger particles. However, according to less restrictive criteria of Agarwal and Liu [1980], the inlet would be more than 90% efficient even for an aerodynamic diameter of 50 μm.

However, experimental results reported by Fairchild et al. [1980], show that at a flowrate of 0.0018 m^3/min., the PM oversamples particles larger than 15 μm in calm air. There is <10% error for smaller particles. At wind speeds of 2.0-5.0 m/sec, the monitor oversamples particles >15 μm when facing the wind, but there is no problem for those <10 μm. At 90° to the wind, the sampler efficiency is decreased. Vincent and Gibson [1981] reported that for some blunt samplers similar to the closed-faced PM, particles that impact onto the face of the sampler may eventually enter the sampling orifice. The net effect is oversampling. Their experiments were done in a wind tunnel with a wind speed of 2.4 m/sec and sampling flowrates of 2 liter/min.

We have measured wind speeds up to 2.5 m/sec at the induction furnace of the beryllium production facility in the summer when large cooling fans are in use. In addition, there are local exhaust hoods close to each furnace. On a cool spring day, however, there was little wind on most of the working platform. In the experiments of Vincent and Gibson, and Fairchild et al., oversampling was most severe when samplers faced directly into the wind. PM are usually at a 90° angle to the wind direction, but the wind may be deflected by the body of the wearer. Thus oversampling may be less serious than reported, but still contribute to the sampling bias.

The size mass distribution of the aerosol must be known to assess the significance of inlet effects. A new adaptation of the multicyclone sampler (MCS) of Lippmann and Kydonieus [1970] was constructed for this purpose. Sample contamination is a serious concern in industrial environments. Size distribution measurements are particularly difficult for Be because of the extremely low TLV (2 $\mu g/m^3$). The MCS permits collection of numerous field samples with ease, and with minimal possibility of contamination. Five 37-mm cassette aerosol monitors, four of which have cyclone pre-collectors, were used in parallel in each MCS. Flowrates were 1.0, 2.0, 3.8 and 4.2 liter/min for the cyclone-filter collectors and 3.9 liter/min for the total dust collector. Each filter was transported to the laboratory for subsequent analysis in its original cassette. The flowrate through each cyclone-filter was controlled with a critical orifice.

The particle size distribution of the Be was measured in calm air with two MCS at the Be-Cu metal alloy casting operation. This area was selected for two reasons: (1) we had some previous data collected at the same site, and (2) the total melting and casting process lasts for about 4.5 hr. One MCS was located about 6 ft from the induction furnace and the second was located about 15 ft from the melt on the operating platform.

Over a two-day period, 16 sets of samples were collected. Sampling periods were selected to correlate with specific operations, and ranged from 31 to 138 min. The filters were wet ashed and Be and Cu were measured by direct-current plasma atomic emission spectrometry (DCPAES). This procedure has been described elsewhere [Chang et al. 1982]. Blank values averaged 0.0047 ± 0.0013 μg/sample for Be (n = 20) and 0.32 ± 0.03 μg/sample for Cu (n = 23). The lower limits of detection (LLD) [Gabriel 1970] are then 0.01 μg/sample for Be and 0.44 μg/sample for Cu.

The Be content of the filter samples was low, but none was below the

LLD. The aerosol was assumed to be log-normally distributed. Typical cumulative frequency distributions are shown in Figure 1. No size distributions were obtained for Cu, because the Cu content of most filter samples that had cyclone precollectors was below the LLD.

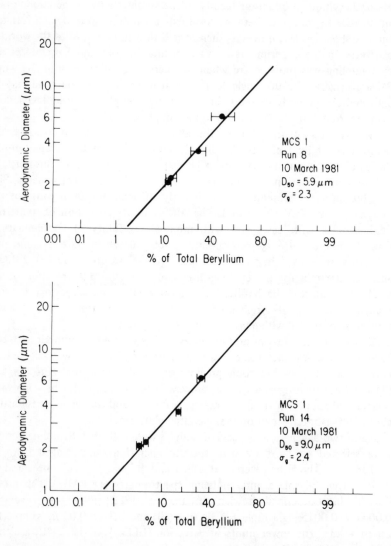

Figure 1. Size mass distribution of Be particles plotted from multicyclone samples collected near a Be-Cu furnace. A log-normal distribution is assumed to estimate the MMAD (D_{50}).

Results of these measurements are summarized in Table I. The mass median aerodynamic diameter (MMAD) ranged 3–6 μm during most of the sampling periods. For two measurement periods, the MMAD was considerably larger (8–16 μ). During both of the latter sampling intervals, the furnace was being "charged," that is, the operator loaded the metal onto the platform and added it to the furnace. As the metal melts and settles into the furnace, additional material is added. Measurements with a Mercer impactor were made at this same site on an earlier field trip. The results were a composite of five sets of samples taken over several furnace cycles because of the small amount of Be that could be collected in one sampling interval. Those results give an overall MMAD = 3.2 μm and σ_g = 10. This is consistent with the present results.

Most of the time, the particles present at the furnace area are in the size range where PM efficiency is near unity for calm air. However, during certain operations, large particles are present and may be oversampled. Because of the low TLV for Be, a few larger Be-containing particles could represent a significant contribution to the total collected over a full shift. Thus, oversampling of the aerosol concentration by PM may account for a portion of the discrepancy reported between the two sampling methods.

It is interesting that Donaldson and Stringer [1980] report total and

Table I. MMAD of Beryllium During Casting of Be-Cu Alloy

Run Number	MMAD (μm)	σ_g	Operation (approximate time)
1	3.3	2.4	Cold start
2	3.5	2.1	(10:00–11:25)
3	8.5	3.7	Initial charge
4	11	5.1	(15:00–17:00)
5	2.5	2.1	Charging; not yet
6	6.1	2.2	molten (17:00–18:00)
7	5.8	2.6	During melt
8	5.9	2.3	(9:30–11:00)
9	4.2	3.8	Complete melt; rub,
10	3.9	2.2	skim, purge (11:00–12:00)
11	4.2	17	Casting
12	5.9	2.2	(12:00–13:00)
13	16	3.5	Initial charge
14	9.0	2.4	(13:00–15:00)
15	3.2	4.1	Complete melt; rub,
16	4.0	2.3	skim, purge (15:00–16:00)

respirable dust concentrations measured with PM in this area to be 2.9 ± 0.4 and 0.9 ± 0.4 $\mu g/m^3$ of Be, respectively. The TWA value is 1.8 ± 0.4. Although the variability is large, our results are consistent with their average respirable fraction of 31%.

There was no correlation between the measured MMAD and either Be or Cu concentration in air. There is an apparent correlation with the ratio Cu:Be in air (Figure 2). The observed distribution is consistent with a simple model for a condensation aerosol with a Be-enriched core and a Cu-enriched surface layer.

FILTER EFFICIENCY

A contribution to the sampling error may result from low collection efficiency of the Whatman 41 filter. Minimum experimental efficiency reported by Stafford and Ettinger [1972] at the face velocity of 0.56 m/sec

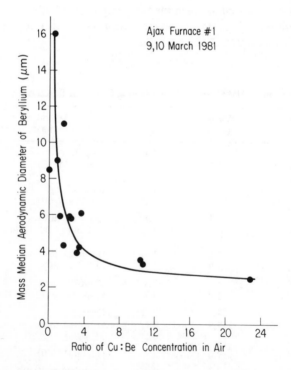

Figure 2.   Correlation of the MMAD of the Be particles near a Be-Cu furnace with the ratio of Cu:Be in air.

is 95% for all particle sizes of DOP aerosol. Some recent measurements indicate that this may be an overestimate for the dry industrial aerosol [Emly 1981].

ELECTROSTATIC EFFECTS

The plastic cassette monitors can develop large electrostatic potentials as a result of friction experienced in normal handling. During the course of some experiments we measured the electrostatic field at 1 cm from the face of the cassette monitors just before sampling. A median electrostatic field of −220 V/cm was measured for 46 monitors (Figure 3). Lovestrand and Rosen [1980] reported that sampling with charged filters can result in greatly enhanced collecting efficiency. Studies of the effect of the electrostatic fields are in progress in our laboratory and will be reported separately.

SELF-DILUTION

Self-dilution by a sampler may occur in two ways. First, samples may be diluted by the return of clean air from a sampler wake to the collecting

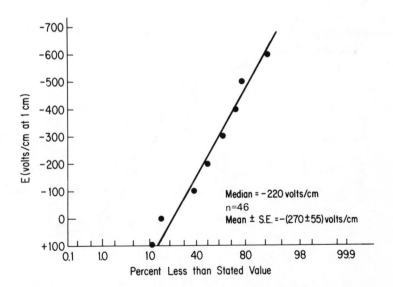

Figure 3.   Distribution of the electrostatic field 1 cm from the face of polystyrene cassette monitors. The fields resulted from normal handling.

volume. Second, air sampled from a contaminated volume will be replaced by surrounding air, which may have a different pollutant concentration. We have carried out a series of experiments in a specially designed dust room to evaluate these factors for the MHV and PM samplers used for Be field measurements. We examined the kinetics in experiments that measured the rate at which the aerosol concentration in the room decreased as a result of sampling. The concentration was monitored with two light scattering nephelometers. Dust was dispersed into the sealed room and, when a concentration of about 20 mg/m³ was achieved, generation ceased and sampling was initiated. The aerosol was thoroughly mixed throughout the experiment and the size distribution was monitored. With this procedure, we were able to calculate what the aerosol concentration would be in the absence of sampling. The exponential decay rate observed as a result of natural removal processes was compared with the rate observed while the samplers were in operation. An example of the concentration changes is shown in Figure 4. It is clear that air cleaning by the MHV must be considered. The effect seen in these tests is amplified by the large volume of air sampled relative to the size of the dust room.

Similar field experiments are not comparable, because we could not deter-

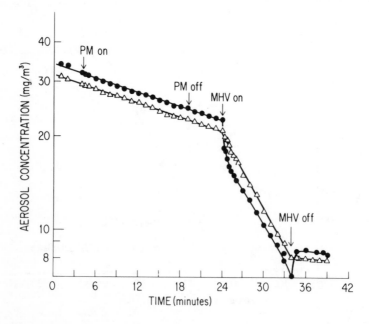

Figure 4.   Reduction of aerosol concentration with time and during sampler operation in a closed dust room.

mine what the aerosol concentration would be in the absence of sampling. To obtain a rough estimate, sequential sampling with the MHV and PM was done at the Be-Cu Ajax furnace area. Aerosol concentration was simultaneously monitored with the nephelometers. We were unable to detect any reduction in the concentration estimated by the MHV as compared with the PM. For these particular experiments, a greater average air concentration was estimated by the MHV. The ratio of concentration estimates MHV:PM was 1.3 ± 0.1. Thus, no effect of air cleaning is observed under these conditions.

VARIABILITY WITHIN THE BREATHING ZONE

Artifacts in exposure estimates may also result from local variation in aerosol concentration within the breathing zone. Location of the sampler can be critical where there is a steep concentration gradient, or if dust is resuspended from clothing. A series of laboratory and field studies were conducted to examine the variability of exposure estimates within the breathing zone.

Baseline data were established in calm air with PM worn at the lapel, nose and forehead. These data were needed to establish whether flow fields around PM would effect the aerosol concentration measurement. Wood and Birkett [1979] and Vincent and Mark [1981] reported that the impacts of the macroscopic and microscopic flow fields around a sampler must be considered, particularly where the flow field is modified by the body of the wearer.

The effects of the position of the monitors was investigated with three well mixed test aerosols [Cohen et al. 1982]. Comparative measurements were made for personnel exposed to a submicron ambient aerosol in the radon calibration chamber at the Environmental Measurements Laboratory (EML) of the Department of Energy (DOE). The radon concentration inside the room is maintained at either 10–15 or 40 pCi/l. The radon gas decays through a series of short-lived radioactive daughters, which rapidly attach to the ambient aerosol, thus providing an aerosol that can be detected readily. The size distribution of the aerosol has been well characterized. The activity median diameter is 0.15 μm. Similar measurements were made on a mannequin exposed to uniform dust concentrations in a specially designed dust room. The mannequin was sprayed with Aerodag G (Acheson Colloid Co.), a dry film conducting material of micron-size graphite to simulate the dielectric properties of the human body. It was dressed in cotton-polyester blend work clothing. The dust room studies were conducted with magnetite (MMAD = 1.6 μm) and Arizona road dust (MMD = 7.5 μm).

The results of these measurements are summarized in Table II. The average mean ratio of concentration measured at the lapel to that at the nose for

Table II. Ratio of Concentration Measured at the Lapel and Forehead
to That Measured at the Nose

Aerosol	Lapel/Nose	Forehead/Nose
Radon Daughters	0.87 ± 0.02	
	1.01 ± 0.02	
	0.95 ± 0.02	
	0.95 ± 0.02	
	1.01 ± 0.02	
	0.98 ± 0.02	0.96 ± 0.02
	0.99 ± 0.02	0.99 ± 0.02
Magnetite	1.02 ± 0.02	0.97 ± 0.02
	1.01 ± 0.02	1.04 ± 0.02
	1.03 ± 0.02	–
	0.89 ± 0.02[a]	0.97 ± 0.02[a]
	1.04 ± 0.02[a]	1.02 ± 0.02[a]
Fine Test Dust		0.98 ± 0.03
	0.80 ± 0.03	0.99 ± 0.03
	1.05 ± 0.03	1.07 ± 0.03
	0.99 ± 0.03	1.04 ± 0.03
	0.95 ± 0.04[a,b]	1.21 ± 0.04[a,b]
	1.08 ± 0.03[a]	1.07 ± 0.03[a]

[a]Aerosol generation took place during part of the sampling period.
[b]No mixing.

all three aerosols was 0.98 ± 0.01 and the forehead to nose ratio was 1.01 ± 0.02. Clearly, for aerosols dispersed uniformly within the breathing zone, under controlled conditions, no bias results from either location when estimating inhalation exposure.

A portable field monitoring system was developed to measure the spatial variability of the aerosol concentrations in the breathing zone of a worker. Three continuous-reading light-scattering aerosol monitors (RAM-1, GCA Corp.) and a tape recorder were incorporated into a specially designed and fabricated backpack. Signals proportional to the dust concentration at the forehead, nose and lapel were recorded on a magnetic tape as a function of time. The sampled aerosol was collected on backup filters for later chemical analysis for Be. The recorded signals were subsequently read out onto a strip chart recorder. Figure 5 is an example of the tracings obtained at the forehead, nose and lapel when the tape-recorded signals are played back into a strip chart recorder. The upper 3 tracings are direct readouts. The lower set are same tape recordings electronically amplified to reveal additional data.

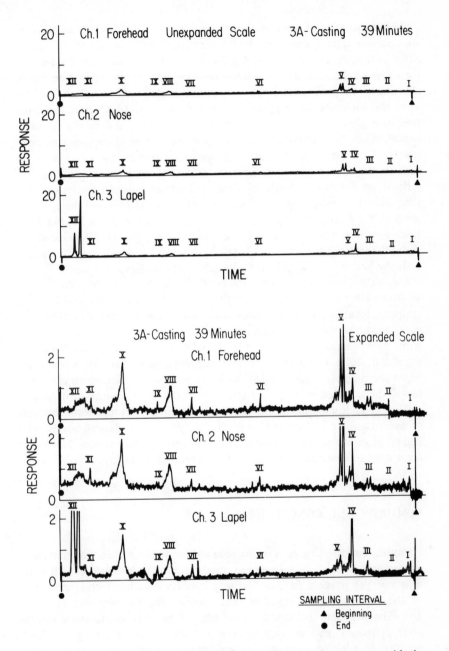

Figure 5. A representative set of tracings of the variation in dust exposure with time of a worker casting a Be-Cu ingot. The simultaneous measurements were made with nephelometers in a backpack worn by the worker.

Field measurements were made of the concentration around the head of an operator during the same Be-Cu metal casting cycle described above. Seven sets of measurements ranging 14–52 min each were made on 5 operators. Five of the measurements were made on the furnace chargeman during charging of the furnace, and two were on the casting headman. Each operator wore the backpack while performing his usual tasks. An observer listed the time and character of every activity of the operator during the sampling period. Thus, it was possible to correlate the measured exposure patterns registered on the magnetic tape with specific activities, e.g., the reason for each numbered peak shown in Figure 5 could be identified.

Based on the best estimate of visual differences between tracings on the strip charts representing the locations sampled, the overall background levels at the forehead are generally greater than those at the other two locations for both the unexpanded and expanded scales. On the unexpanded scale, the levels at the lapel appear to be the lowest, whereas on the expanded scale the lowest levels appear as frequently at the nose as at the lapel.

The strip chart curves were integrated with the graphics capability of an Apple II+ computer to estimate the average dust concentrations for each sampling location (forehead, nose and lapel). The relative measured dust concentrations were very variable. The data are represented in Figure 6 with 95% confidence limits. The correlation between unexpanded and expanded scales is better for higher concentrations than for lower ones. When the monitor is operating at the lower limit of the selected measurement range, the correlation is less satisfactory. The highest average dust concentration in each experiment resulted from either a short high-level exposure or overall greater background levels. No body locations showed consistently higher or lower average measured dust concentration values. More data are needed to apportion the source of the observed variability.

SUMMARY AND CONCLUSIONS

Factors contributing to the discrepancy between estimates of inhalation exposure to Be by different sampling methods have been identified. Although it is not yet possible to quantitatively apportion each effect as a fraction of the total disparity, together they account for the observed factor of 2–3. The variables include sampling bias of the PM, electrostatic effects, filter efficiency and dilution of the aerosol concentration by the MHV sampler. In addition, because lapel-mounted monitors sample local air, the variability of the concentrations within the breathing zone, particularly as a result of resuspended dust, is an important factor.

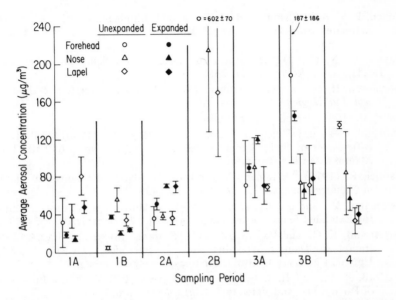

Figure 6. Aerosol concentration at the forehead, nose and lapel of a worker averaged over an entire sampling period. Results are shown for seven sets of measurements for direct signal transcript (unexpanded) and for the amplified (expanded) signal.

ACKNOWLEDGMENTS

This research was supported by a contract from Brush Wellman Inc. and is part of a center program supported by the National Institute of Environmental Health Sciences, Grant No. ES 00260. The authors wish to thank Dr. Mark Emly for his collaborative efforts during the field studies, and Mr. Robert Morse, who performed the DCPAES measurements.

REFERENCES

Agarwal, J. K., and B. Y. H. Liu (1980) "A Criterion for Accurate Aerosol Sampling in Calm Air," *Am. Ind. Hygiene Assoc. J.* 41:191.

Chang, A. E., R. Morse, N. H. Harley, M. Lippmann and B. S. Cohen (1982) "Atomic Emission Spectrometry of Trace Levels of Beryllium in Industrial Aerosols," *Am. Ind. Hygiene Assoc. J.* 43:117.

Cohen, B. S., A. E. Chang, N. H. Harley, and M. Lippmann (1982) "Exposure Estimates from Personal Lapel Monitors," *Am. Ind. Hygiene Assoc. J.* Vol. 43.

Dayies, C. N. (1968) "The Entry of Aerosols into Sampling Tubes and Heads," *Staub-Reinhalt. Luft* 28:1.

Donaldson, H. M., and W. T. Stringer (1980) "Beryllium Sampling Methods," *Am. Ind. Hygiene Assoc. J.* 41:85.

Emly, M. (1981) Personal communication.

Fairchild, C. I., M. I. Tillery, J. P. Smith and F. O. Valdez (1980) "Collection Efficiencies of Field Sampling Cassettes," Report No. 8-640-MS, Los Alamos Scientific Laboratory.

Gabriel, R. (1970) "A General Method for Calculating the Detection Limit in Chemical Analysis," *Anal. Chem.* 42:1439.

Lippmann, M., and A. Kydonieus (1970) "A Multi-stage Aerosol Sampler for Extended Sampling Intervals," *Am. Ind. Hygiene Assoc. J.* 31:730.

Lovestrand, K. G., and V. Rosen (1980) "Storande Invekan av Statisk Electricitet vid Dammprovtagning Med Filtermetoder," UURIE 136-80, Uppsala University Institutet for Hogspanningsforskning.

Stafford, R. B., and H. J. Ettinger (1972) "Filter Efficiency as a Function of Particle Size and Velocity," *Atmos. Environ.* 6:353.

Vincent, J. H., and H. Gibson (1981) "Sampling Errors in Blunt Dust Samplers Arising from External Wall Loss Effluents," *Atmos. Environ.* 15:703.

Vincent, J. H., and D. Mark (1981) "The Bases of Dust Sampling in Occupational Hygiene: A Critical Review," *Ann. Occup. Hygiene* 24:375.

Wood, J. D., and J. L. Birkett (1979) "External Airflow Effects on Personal Sampling," *Ann Occup. Hygiene* 22:299.

AUTHOR INDEX: VOLUME 1

Page numbers in boldface indicate authors of chapters in this book.